Praise for *The Ethical Meat Handb*

This book is practical, yes, but it's also deeply personal. Meredith Leigh will teach you how to raise animals, butcher them, and cook and cure their meat. Even better, she explains what it means and why it matters, and her passion is infectious. After reading this book, I longed to smell the deep funk of the barn, to feel the squish of mud beneath my chore boots, to heft a butcher knife in my hand.

—Mark Essig, author of *Lesser Beasts: A Snout-to-Tail History of the Humble Pig*

This is a powerful, positive book about a powerful, positive alternative, engaging us in shaping a new food and agriculture narrative. It leaves us with an overwhelmingly clear picture of just how important ethical meat is to the future of the entire food system. It is exactly what our movement needs, a call for people to take action, educate themselves, and participate. This book is a gem, a gift to everyone looking to be the alternative to a passive recipient of the status quo. This book— its philosophy, honesty, artistry, and practice, represents a way forward — the only way to go.

—Jean-Martin Fortier, from the Foreword

With a warm, welcoming tone, Meredith Leigh invites readers to the table to learn everything there is to know about raising, slaughtering, butchering, and preparing good meat. This is the most comprehensive guide that I have ever seen on the subject. I'm looking forward to trying some of the delicious looking Southern-inspired recipes in the book too. Ms. Leigh has obviously put a great deal of work into this book and it shows.

— Rebecca Thistlethwaite, Farm & Sustainability Consultant and author, *Farms with a Future: Creating and Growing a Sustainable Farm Business*, and *The New Livestock Farmer: The Business of Raising and Selling Ethical Meat*

the Ethical Meat handbook

complete home butchery, charcuterie and cooking for the conscious omnivore

Meredith Leigh

Cover design by Diane McIntosh. Cover illustration by Rob Hunt.

Printed in Canada. First printing August 2015.

New Society Publishers acknowledges the financial support of the Government of Canada through the Canada Book Fund (CBF) for our publishing activities.

Any other inquiries can be directed by mail to:

New Society Publishers
P.O. Box 189, Gabriola Island, BC V0R 1X0, Canada
(250) 247-9737

LIBRARY AND ARCHIVES CANADA CATALOGUING IN PUBLICATION

Leigh, Meredith, 1983–, author
The ethical meat handbook : complete home butchery, charcuterie and cooking for the conscious omnivore / Meredith Leigh.

Includes index.
Issued in print and electronic formats.
ISBN 978-0-86571-792-3 (paperback). — ISBN 978-1-55092-603-3 (ebook)

1. Cooking (Meat). 2. Meat—Moral and ethical aspects. 3. Meat industry and trade—Moral and ethical aspects. 4. Slaughtering and slaughter-houses—Moral and ethical aspects. 5. Cookbooks. I. Title.

TX749.L44 2015 641.6'6 C2015-905547-4
C2015-905548-2

New Society Publishers' mission is to publish books that contribute in fundamental ways to building an ecologically sustainable and just society, and to do so with the least possible impact on the environment, in a manner that models this vision. We are committed to doing this not just through education, but through action. The interior pages of our bound books are printed on Forest Stewardship Council®-registered acid-free paper that is 100% post-consumer recycled (100% old growth forest-free), processed chlorine-free, and printed with vegetable-based, low-VOC inks, with covers produced using FSC®-registered stock. New Society also works to reduce its carbon footprint, and purchases carbon offsets based on an annual audit to ensure a carbon neutral footprint. For further information, or to browse our full list of books and purchase securely, visit our website at: www.newsociety.com.

Contents

Foreword... *by Jean-Martin Fortier* . . . ix
Prologue: Why You Should (or Shouldn't) Read This Book . . . xiii

Introduction: The Ethical Meat-Eater . . . 1
What is Ethical Meat? . . . 1
Buying Differently . . . 4
Cooking Differently . . . 12
Eating Different Things . . . 15

General Notes on Raising, Cooking and Eating Animals . . . 25
Slaughter . . . 25
Notes on Cooking and Eating Muscle . . . 26
Disclaimers . . . 27
Butchery Tools and Tips . . . 28

1. Beef . . . 35
Raising Beef . . . 36
Breeds . . . 37
Space and Water . . . 42
Fencing . . . 43
Feed and Minerals . . . 44
Beef Butchery . . . 48
Cooking with Beef . . . 75
Beef Stock . . . 76
Beef and Lovage Sausage . . . 77
Braised Beef Shank Tacos with Caper Chimichurri . . . 78
Beef Bacon . . . 79

Beef Tallow . . . 79
Beef Jerky . . . 80
Bresaola . . . 80
Sauces and Sundries for Beef . . . 81

2. Lamb . . . 83
Raising Lamb . . . 85
Breeds . . . 88
Space and Water . . . 90
Fencing . . . 91
Feed and Minerals . . . 92
Lamb Butchery . . . 95
Cooking with Lamb . . . 109
Earl Grey Braised Lamb Shank with Herb Dumplings . . . 110
Lime-Cream Curry Lamb Sausage with Dosas and Raita . . . 111
Fire-Cooked Lambchetta with Apricot and Rosemary . . . 112
Roasted Lamb Rib with Orange, Fennel and Honey Marmalade . . . 113
Roast Leg of Lamb with Mustard, Capers and Marjoram . . . 114
Bourbon and Sorghum Glazed Lamb Spare Ribs . . . 114
Sauces and Sundries for Lamb . . . 115

3. Pork . . . 117
Raising Pigs . . . 120
Breeds . . . 122
Space and Water . . . 123
Fencing . . . 127
Feed and Minerals . . . 127
Pork Butchery . . . 128
Cooking with Pork . . . 145
Pulled Pork with Hot Vinegar Sauce, Chow Chow and Corn Pancakes . . . 147
Pork Banh-Mi Sandwiches with Quick Pickles . . . 148
Braised Pork Ribs with Rooster Sauce and Balsamic . . . 149
Chicharron with Apple Butter and Cilantro Crème Fraîche . . . 150
Lard . . . 151
Basic Pie Crust . . . 151

Pork and Pickle Pie . . . 152
Breakfast Scrapple with Arugula, Eggs and Maple . . . 153
Porchetta with Persimmon, Chestnut and Pine . . . 154
Sauces and Sundries for Pork . . . 155

4. Charcuterie . . . 157
Deciphering Cured Meats . . . 157
Getting Started with Fresh Sausage . . . 164
Breakfast Sausage . . . 168
Chorizo . . . 168
Herbes de Provence Sausages . . . 169
Garlic Orange Bratwurst . . . 169
Pates, Terrines, and Meat Specialties . . . 170
Liver Pâté . . . 173
Quatre Epices . . . 174
Headcheese . . . 175
Bologna . . . 176
Whole-Muscle Cures . . . 176
Master Dry Cure Recipe . . . 177
Bacon . . . 178
Smoking Meats . . . 178
Pancetta Stesa . . . 179
Prosciutto . . . 180
Capicola . . . 181
Lardo . . . 181
Smoked Fiochetto Ham . . . 182
Fermented Sausages . . . 182
Basic Salami . . . 186
Fennel Salami with Nutmeg and Wine . . . 187
Pepperoni . . . 187

5. Poultry . . . 191
Raising Poultry . . . 194
Breeds . . . 197
Housing and Fencing . . . 200
Feed, Minerals and Water . . . 202
Additional Considerations . . . 204

Home Slaughter 206
Chicken Butchery 211
Cooking with Poultry 218
Spatchcocked Roasted Chicken with Lemon and Basil 218
Fried Chicken 219
Chicken Ballotine, Three Ways 220
Duck Confit 221
Duck Rillettes 222
Chicken Cardamom Sausage 223
Sauces and Sundries for Poultry 224

Resources 225
Thanks and Praise 229
Index 231
About the Author 239

Foreword

by Jean-Martin Fortier

A long time ago I read a book called *The New Organic Grower,* which opened my mind to the importance and practice of small-scale vegetable production. Eliot Coleman's famed book not only gave instructions about how to do things, it provided an ethos about good food and the way forward. This book changed my life and shaped the path I've been following ever since. Books such as his are just as rare as they are important, because they can transform ideas of sustainable living into practical and proactive practices. I believe *The Ethical Meat Handbook* is one of these books.

As a vegetable grower, my focus has always been on the plant side of things. My farm is nothing more than a big garden, and if you read some of my work, I've often advocated that young growers NOT include animals in the system, arguing that animal husbandry could enslave you to your farm and require a lot of initial investment.

Lately, I've been changing my mind about that, mostly as I've learned more about the management of perennial systems, based on pastured land where herds of animals are moved very frequently. These are the low-tech, high-management systems that my friend Joel Salatin has been teaching us about for nearly 30 years, systems that don't require land to be tilled, or for much grain to be fed to livestock.

As I have learned more about polycultures and the overall importance of animals in a restorative farming model, I have had a joyful change of heart.

The fact that every good farmer is aware of, regardless of his conscious practice, is that animals and plants managed together, holistically, provide powerful ecosystem benefits via natural trophic exchange.

Plants are the only beings that can turn solar energy into usable energy for all life, but they depend on high energy being returned to the soil. As such, animals offer a simple, closed-circle fertility input that is beneficial to our farmland. When managed properly, and holistically, animals and plants together provide sustainable solutions.

As I have begun to learn about holistic animal agriculture, and work to merge it with my current farming consciousness, one issue has remained to nag at me: what about the slaughtering and butchering of the livestock? Where is the system or thinking for post-production that feels right, and ready? And it is here, at the intersection of the right production points, and this nagging question, that *The Ethical Meat Handbook* soars. It brings to light not only an ethos about meat consumption that I believe we should all agree on (both meat eaters and vegetarians alike), but most importantly, it also provides guidelines about how to conduct ourselves responsibly in relationship to the earth.

By drawing on hard-earned experience and profound insight, author Meredith Leigh challenges us to consider ethical meat production as center stage in our food conundrums; or at the very least, recognize the threads in our fraught and flawed meat industry that are patterned through our entire food system. Further, she challenges us to see that a proliferation of small-scale butchers working in relationship with local farmers will play an important role in the rebuilding of our food systems. The evidence of this is clear: People strongly opposed, saddened and angry at the way our society treats its animals in factory farms are looking for alternatives, and they can find it in ethical meat, locally raised and butchered. The demand for locally grown products is driving a new cohort of young people back into the business of small-scale agriculture, and this demand will only move forward as we increase our understanding of the ecological, economic, and social importance of ethical meat.

An improved food system will depend on independent, resilient, integrative small businesses, all fostered by a renewed and growing interest in local agriculture. From small farms to feed mills, farmers markets to bakeries, this is the middle economy that *The Ethical Meat*

Handbook so clearly illustrates. Where livestock is used in combination with plant agriculture to create healthy food, fiber, and ecological management, the artisan butcher will be indispensible. It's a trend that will inevitably create plenty of opportunities for people interested in pursuing farming careers and lifestyles.

This is a powerful, positive book about a powerful, positive alternative, engaging us in shaping a new food and agriculture narrative. It leaves us with an overwhelmingly clear picture of just how important ethical meat is to the future of the entire food system. It is exactly what our movement needs, a call for people to take action, educate themselves, and participate. This book is a gem, a gift to everyone looking to be the alternative to a passive recipient of the status quo. This book—its philosophy, honesty, artistry, and practice, represents a way forward—the only way to go.

JEAN-MARTIN FORTIER is a small-scale organic farmer, writer, and educator. He and his wife are the founders of Les Jardins de la Grelinette, an internationally recognized micro-farm famous for its high productivity and profitability using low-tech, high-yield methods of production. His acclaimed book, *The Market Gardener: A Successful Grower's Guide to Small-Scale Organic Farming,* tells the story of how they started their farm, and how they successfully operate it. [themarketgardener.com]

To Big Dipper and Little Dipper,
my boys, *mes petites*.
Someday you'll know
what incredible good you made of me.

And of course:
To all the brave farmers and
all the sweet beasts.

PROLOGUE

Why You Should (or Shouldn't) Read This Book

I love food. It is constantly on my mind. Granted, food is my work, but even if it wasn't I think I would still be buying flying saucer squash at the market even when my garden is overflowing with lemon squash, well, just because they are beautiful and shapely. Food is so fascinating. From how it is grown to how it affects us nutritionally to how it is cooked and preserved, there are more than a million possibilities for thought and practice. Is it not delightful that when a predator bug lands on the leaf of a brussels sprouts plant, the plant recognizes the insect and begins to release hormones that make its leaf less palatable to the enemy? Isn't it humbling to know that without a hundred trillion tiny bacteria in our intestines, we would not be able to maintain healthy digestion? How intriguing that purple beans turn green when cooked, and milk tastes different depending on what grasses the cow ate, and pork shoulder contains an almost perfect ratio of lean meat to fat for making sausage. This book is about meat, yes, but it's about honest integration of meat consumption into a robust, diverse diet. Ultimately, it is about food, because good meat is made better with other foods: great herbs, good salt, fresh vegetables, or phenomenal sauces, to name a few.

Since I've given up commercial farming (for now) and working in a commercial kitchen (for now), I've moved to town and started living life on this *other* side of food, this post-production angle. It's giving me a wonderful new perspective. I shop every Saturday morning at the North Asheville tailgate market in Asheville, North Carolina. I stock up for the week and also buy extra food to preserve indefinitely. At the market, there is this one booth that has prevailed over all others as my favorite: Anne Grier, at Gaining Ground Farm. I have such respect for this woman's food. For weeks now, I have been trying to figure out why. I respect all the food at the market—food that is whole and fresh. I love so many of these farmers and support them all with my money and my words. But this Gaining Ground food has me riveted, every week. What a peculiar feat! This is good, good food. Let's figure out why.

What is good food? It is not entirely subjective. The fact that something might taste good to me and not to you has something do with cultural perspective, or each of our unique experiences with taste and texture in general. With respect to these factors, goodness is subjective as sure as is rottenness. But that subjectivity has only a fraction to do with the food's actual worth. There is also universally good food, and masterful cooks understand how to demonstrate its worth to both the discerning and the under-educated eater.

Food that is universally considered good is almost always fresh food, well raised, from superior resources. Let's take bacon. Your experience is richer with bacon you've made yourself—possibly with a pork belly that you cut from a pig you raised who rummaged for mast and grub in dark soil and forest litter—than it is when you breeze past the pale, vacuum-sealed bacon at the supermarket, from a distant pig brought up in a concrete pen. We can deduce that food is better when the earth that grows it is healthy, and its source and fabrication are closer to you. If you're reading this book, you probably already accept this. Easy, you say. Gaining Ground's vegetables are better because they are organic and local.

But, wait. Gaining Ground Farm's food is no more local or organic than the other farmers' fare. Why is it that I somehow perceive it as different? Is it Anne's presentation? Is it her perspective? Come to think of it, I bought the bunch of marjoram from her just because she harvested it when the flowers were swelling and about to bloom. It was fragrant—her entire booth smelled like soup, what with her putting the herbs right under the celery. Did she do that on purpose? I picked up the marjoram and put it down, thinking *I don't need this*. But then I picked it back up when I realized that I absolutely *needed* to pull every last flower bud off of the marjoram and fry them, in butter and flour and panko, for my fava bean and beet salad. *Done. Into the bag.*

The colors at Anne's booth are stunning. The brown-red of the potatoes, their papery skin rubbed off in places from handling, to reveal their white flesh. The greens of zucchinis, both the forest green skin of raven and the ribbed, kelly green and mint skin of costata romanesco. The yellow squash, the golden and burgundy beets, the purple kohlrabi. There is not a lot of packaging, either. Everything spills from boxes and piles on the table. I must touch every piece to claim it for myself, bagging it and taking it to Anne for weighing. And oh, the bounty. Everything is absolutely overflowing and speaking of the earth that grew it.

This is it. Good food ignites all of our senses. We interact with it. Truly good food goes so far as to improve our sensory experience, so that we can teach ourselves how to appreciate it. When we experience really good bacon, for example, our eyes see beautiful striations of velvet-white fat and deep red lean; our mouths water as we imagine the taste; our ears hear the crack of fat in the pan; and our noses twitch at the smell of smoke and salt and pork. What we taste is history, as well as hickory, the inevitable morning mist on a farm, the memory of fire. What we taste, if it is really good, will be recorded for later, and not just filed into our intestines and shit out before bed.

If your food is to be good, it must be an experience, a sensory journey that can drop your guard. The experience of good food is not just resigned to taste. Even excellent cooks do not succeed every time. The key is to honor the process, the pursuit of good food. Don't be afraid, and definitely do not settle. You should read this book to inspire yourself to a point of spirituality toward the animals you raise, the feeling on your hands as you rub salt into a hunk of meat, the few moments you take to consider the best herb for your sauce, and the inspiration and ambition you feel as you wait for a cure to set. Read this book as part of your worship of food—for the experience, for what you will become because of it, above and beyond the nourishment it provides. For the community that will come to your table to partake, for the ideas you come up with for next time.

You need food. What everyone will not admit is that you should love it, too.

If you don't love food, put this book down. Now.

The Ethical Meat-Eater

What is ethical meat?

- **Ethical meat comes from an animal that enjoyed a *good life*.** The animal acted out its natural tendencies, in a way that did not over-deplete resources but contributed to healthy natural cycles. It was cared for and not neglected. It endured little stress.
- **Ethical meat comes from an animal that was afforded a *good death*.** The animal endured little stress in handling on its way to slaughter. It did not suffer long, but was slaughtered in a way that rendered it unconscious instantly, and then humanely relieved of its blood.
- **Ethical meat is *butchered properly*,** making full use of the carcass out of thriftiness, efficiency and respect for the life that was given as food.
- **Ethical meat is *cooked or preserved properly*,** maximizing nutritional benefit and paying homage to the important rituals of deliciousness.

I was a vegetarian for nine years, and a vegan for two. I watched a grueling video in high school about the horrors of an industrial slaughterhouse. I did some light reading in environmental philosophy, and made a decision. I was largely ignorant. I was not making a huge difference in the lives and deaths of animals, was not looking at the bigger picture of global human health and environmental restoration, was not actively changing mass wrongdoing. I was motivated by deep empathy and justified political aggravation, but my solution, sadly, mostly helped only me.

I spent my college career learning what I could about the scientific, political and cultural intricacies of agriculture. I traveled to different countries, learned about drastically different attitudes toward food and land, and saw the ways that people have shaped their corners of the earth in the quest for nutrition. In Vietnam it is a gesture of friendship to place food in another's bowl. When, in 2004 in a rural Hai Duong village in northern Vietnam, a small woman named Loi placed a stringy piece of water buffalo into my dish at dinner, I began my journey into the meaningful consumption of animals.

Before that moment, my diet had been one of luxury, and a desire to escape a system I felt I could not affect. When I ate the piece of flesh as an act of communion, I checked in to another way of thinking. Eating gained new meaning, as I was very aware that Loi herself had milked and cared for, and eventually slaughtered, that animal for our meal. I started to look for the bigger picture, and solidified my decision to devote my life to food. I have spent the decade since as an omnivore, working with food from almost every angle, with the belief that we can make a difference in the well-being of plants, animals and the earth, while still loving all food and seeking good health.

Time and again in North America, we're handed myriad reasons to question our food supply. Between climactic pressures, environmental resource limitations, food safety scares, political maneuvering, media hullabaloo, corporate mergers, impending energy crises, trade deals, population woes, consumption rates, worldwide hunger and poverty, and dominion over the very seed required to create the next generation of food and fiber, we're constantly vacillating, with our big national voice, between justification and condemnation of a globalized food system. Within this passion play, consumers, with their tiny individual voices, have both ultimate power and very little power at all. We drive the machine with our buying dollars, but we are simultaneously so hoodwinked by marketing ploys, dietary "rules" and nutrition trends that we become overwhelmed, dependent and easily duped.

Within this maelstrom, the meat and dairy sector are continually at the eye of the storm. Meat has been demonized since the 1960s, when our nation became afraid of fat and cholesterol. Since then, depending on what research we favor, meat and dairy are either entirely responsible or completely forgiven for all our health woes. Regardless of the trending attitude toward saturated fat, animal protein and cholesterol, we find it easy to banish animal products from our diet when we hear about inhumane treatment of animals, confined animal feed operations (CAFOs), pink slime in ground beef and the effects of added hormones and antibiotics on our meat. Yet I haven't set out to write a book revealing the horrors of the industrial food system or the meat industry within it, and I certainly do not aim to defend either. Others have done plenty of this work already, on both sides. Instead, this book seeks to offer alternatives to the status quo. It seeks to educate buyers and homesteaders about their role within the whole. In other words, you are not a victim; you are not helpless; and you are not merely the last link in a long chain of missteps, bloodlust and greed.

I join eaters everywhere in their opposition to genetically modified organisms, overuse of antibiotics and inhumane living conditions for all beings. I also seek to understand the vast lattice of past and current political, social, economic and environmental factors that make the question of what to put in your mouth three to six times a day very perplexing and outrageous indeed, whether you choose to eat meat or not. Our oppositions can be simple and absolute. Our options are not so easy. This book asks a number of "how" questions, and offers deeply pondered possible answers. How can we work from within a fantastically flawed food system to create real food? How can we work in accessible ways, without alienating any food citizen or farmer? How is it possible to create models that drive an economy, social synergy and environmental restoration that work for the world as we know it *now, and* the world we want later?

I urge you to come by your food more honestly by exploring the ideas presented in this book, because I believe there is a lot more the everyday food citizens can change—and that we can eat a lot better in the process, too. May we endeavor to source and consume meat with more of an understanding of the issues across the supply chain; checking out is not our only option. I'd argue, too, that it's not the best option. Nor is it viable to make more and more demands of farmers, regardless of the size and type of their farms. If you come away from this book with nothing but a sausage recipe and one fun fact, let that fact be this: Across the meat supply chain, the farmer makes the least amount of money, and has possibly the most difficult and sacred job in the journey. It's time to kick it up another notch and realize that truly ethical meat is going to take community effort. If we are to be ethical meat eaters, or good eaters at all, we will buy differently, cook differently and eat different things.

Buying Differently

We cannot expect ethical meat, or any other truly better food, to simply arise in some pure form from the food system we currently have. If we don't change the system, we will constantly be required to compromise what we know is right, and euphemize what we know is happening. Our behemoth of a food industry, which supports suffering and whole system degradation of epic proportions, is often justified by asserting that it is our job to feed the world. This goes with an unspoken assumption that there is only one way to feed the world, and it must be the way that we've found, and we must be doing it now. I'll not surprise you by saying that we are not, in fact, "feeding the world"—and there is another way. Whenever possible, food and necessity should happen in a sphere close to home, in an economy of body and household that I call the "first economy." After that, food should happen on the community level, in systems I call "middle economies." It is possible to foster agriculture on every soil that can feed communities, and so for basic needs, and that vital sovereignty for all, functioning middle economies are a more hopeful way to feed the world. Instead, a whole range of factors, mostly driven by money, have led us askew, into a dependency on "external economy," a vast system that takes place far away from us and involves too many players and too many resources.

I have been involved in movements for nearly fifteen years that address this fundamental issue: how to create middle systems within the current status quo, to produce food, and more. Many days, I wonder if it is working, and it is difficult to imagine us ever entirely abandoning external economy, because it now stretches across the globe. But I have seen small farms, conscious eaters and effective activism grow exponentially over the years, which leads me to think that we must keep trying. And like it or not, the huge and dysfunctional external system is what we are working with *right now*. We cannot avoid it, and we are both contributors to it and victims of it, both farmers and non-farmers. Within this reality, we need to promote first and middle systems as much as possible, because they are smaller, more synergistic, and include plants, humans and non-human animals, from which arise more conscious economies and trade. We also need to try to apply the positive aspects of synergy and diversity to larger systems, to see if that works as well.

As a result of the system in place today, even if your meat has been fed organic grain, it may not have lived well. Even if your meat has suffered less in life, it may not have died a just and clean death. Some of the opportunities you have toward truly good meat come from extremely enterprising, well-meaning, expensive and risky capital investments in good farming; these efforts deserve our every praise, even if we are still struggling to see them grow to accommodate our needs. But other opportunities toward good meat come from extremely enterprising efforts to capitalize only on your desire for good meat; these efforts deserve our every skepticism. Unfortunately, the honest food citizen, with troubles of her own, living her amazing and busy life, is hard pressed to know if she is facing a praiseworthy effort toward good food, or a backward and greedy one. This is the catch-22 of our attempt to repair whole-scale foodthink, figuring it out as we go. This book does not pretend to have answers. Instead, it simply seeks to honestly air the conundrums, show us that we have more in common than we think, and assert that it is worth it to *keep trying* different agriculture, different economy and different philosophy to *improve all life*.

I believe that, *right now*, the best way to access good food and good meat is by raising it ourselves, or by buying it directly from a fellow community member who has done so. And for those of us not able or willing to produce our own animals for meat, I assert that exceptional, good meat, *right now*, for all who endeavor to support it, will require us to pay more money, stimulating a "middle market." This is not an obvious activism for most people, because our current system supports tricky, big, global economics, whether we are buying meat by the carcass or by the cut-and-ready steak. As a result, we constantly buy meat from local farms like it is meat from industrial farms, and it is not the same thing.

Meat costs more than you realize. So does all food. If you're purchasing from the supermarket, you're buying meat that is heavily subsidized by the government (via your tax dollars), a process that removes much of the risk and cost of its production and allows the industry to drive a competitive price at the point of sale. Additionally, much of the meat from larger farms comes from vertically integrated food businesses, meaning that the business owns more than one piece of the supply chain, and thus decreases its cost.

Let's take chicken, for example. A vertically integrated poultry business owns the hatchery (baby chick farm), the chickens (via contracts with farmers), the slaughterhouse and the entire packaging and distribution infrastructure. The corporation owns the whole process, from egg to table. This benefits the company because as the product changes it becomes more valuable, and the ensuing profits stay within the company. The costs of taking the product through all these changes are decreased, because there are not three or four different companies along the way trying to eke profit out of their rung on the ladder. And waste and cost can be controlled and even offset by the company anywhere along the way.

In his book *Meatonomics*, David Robinson Simon cites the many issues within the meat economy, especially the role of farm subsidies. He claims that industrial hog farmers pay an average of eight dollars more than they make on each animal to raise it, and that corporate beef producers spend twenty to ninety dollars more than each animal is worth to raise cattle. While Simon ignores a few complexities within the meat industry, lumping players together and branding all producers as conspirators against the hungry but innocent taxpayer, I find his number crunching on production costs and consumer price perceptions valuable.

How is this backward economy possible? Due to subsidies, which incentivize the growers to continue producing; due to vertical integration, which allows the meat businesses to make that money back as the product moves up the supply chain; and finally, due to the sheer size of the operations. The more chicken our sample corporation offers the market, from whole birds to bone-in thighs to emulsified cartilage and string meat for nuggets, the less the company needs to charge on each product before breaking even.

Instead of paying the true cost for your food at the point of sale, you're currently paying for it in pieces. And the more corners that are cut in its production, the more you pay later, in higher healthcare costs and in degradation of your environment. If we can begin to see, as Simon asserts, that a Big Mac, which normally retails for about four dollars, should really be selling for about eleven, we can begin to see how strapped a small, family farmer must be. I know from experience. I began my journey into what would amount to a decade of farming in 2003, growing organic vegetables, cut flowers and meat. We raised a diversity of crops and livestock on our farm, to increase our marketing appeal, maximize nutrient cycling on our land and feed our family. We were doing what many small farmers feel called to do: creating a middle market for meat and other food that people could trust, and that could stimulate the local economy.

We had mixed results making money as community-supported farmers. Both my husband and I worked off-farm jobs full-time, while raising a family and trying to build, manage and market a farm on a big enough economic scale to support ourselves. We faced many problems in all of our enterprises, headlined by production inefficiencies and economy-of-scale issues. We had neither large enough numbers of animals on the ground, nor the systems in place to raise animals in large enough numbers. When it came to the production of animals, we faced these main obstacles:

1. The high cost of *feed* inputs, largely not customizable by us *in relation to the price we could charge at the point of sale.*
2. The high cost of *slaughter and processing,* largely not customizable by us due to regulatory obstacles, and *in relation to the price we could charge at the point of sale.*
3. The *growing demand for gourmet and niche meat products* such as heritage breed, certified-organic/non-GMO and further-processed foods that we were inconsistent in our ability to profitably produce, due to reasons #1 & #2 above, as well as the price and volume competition we faced from vertically integrated industrial agribusiness. (See the sidebar case study: The True Cost of Organic, GMO-free Pork, p. 20).

Note the threads in these issues, which one might call roots. One is the *lack of control,* resulting in the need to outsource parts of the operation, which always costs money and always limits quality options. The other is the *disconnect at the point of sale.* Our middle business, operating within and alongside the huge, external system madness that our customers also patronized, made it very difficult for us to garner middle market prices. Community-level middle economies face these deeply rooted issues every minute of every day.

Our solution to the problems we faced was an attempt to specialize. We dropped commercial production of vegetables completely, drastically downsized commercial cut flowers, and focused on meat. We scaled up the number of animals, paid closer attention to feed and breeds, and sought to educate our customers and our processor about the difficulties within the supply chain that put limitations on the end product. We began carrying specialty products such as rubs and dry-cured salamis, produced by others with niche meat business ventures. At the end of 2012, when we saw that we were paying 52 percent of our gross profits to processing and packaging, we developed plans for our own kind of middle-system vertical integration: a butcher shop.

The butcher shop would allow us to pay the processor only a kill fee, to slaughter and dress (remove the innards from) the animals. Then we could butcher the animals further at our own facility, turning them into retail muscle cuts, specialty fresh sausages, cured and smoked meats and other items. This would vastly increase the diversity and number of products available to our customers, and put processing revenues into our pockets, rather than someone else's. Granted, we would have additional labor and overhead costs establishing the shop, but our projections showed potential.

The shop opened in October of 2013, and improved our processing costs and product availability to our customers almost immediately. But the farm still faced the issue of feeding animals on a large scale. Even with the shop buying animals at a reasonable price per pound on carcass weight (the weight once the animal is killed and dressed), the farm remained financially stressed. Then our marriage very suddenly collapsed. The farm shut down. All the animals were sold. The shop remains open, and maintains tight margins as it tries to offer only local meat to the communities in and around Asheville, NC.

I firmly believe that a farm such as ours, with a sister business like our shop, can become a viable model for food entrepreneurs in the growing middle food economy. But it will take community effort. It will take increased mindfulness and ingenuity among both farmers and consumers. The fact remains that farmers are still facing the same problems mentioned above, while trying to create new paradigms within the current framework. To boot, not every meat farmer can open a butcher shop, and in many ways, farmers are in the same boat as consumers, in terms of what type of food and agriculture they can afford to throw their weight behind. Ultimately, the consumer faces these problems as well, as he or she asks for reasonable prices on the finished product and seeks the cleanest product possible. More often than not, the consumer is unaware of the premium he or she is asking for, and what business, profitable or not, is behind each premium applied as the meat travels along the supply chain.

So next time you see that sign for boneless pork chops at $6.99/lb at the local grocery, ask yourself where that meat came from, and what systems are in place on a massive scale to drive that price. And then please, do not proceed to the farmers' market and ask why the boneless chops there cost $8.99/lb when you can get them at the local grocery for $6.99. When you make this argument, you are basically asking a farmer why you can't pay commodity prices for a homegrown pork chop—raised by a non-vertically integrated family business, who re-

Enterprising in Agriculture

If you are the investing type, I urge you to invest in farmers. Land may be a place where people put their money, but land-based business is generally hard up. From my own experience as a grower, I know what obscene risks and stresses I undertook in order to try better farming practices, especially on a scale that could function alongside and within our current food system. While researching this book, I caught up with Jamie and Amy Ager, the owners of Hickory Nut Gap Farm (HNG), in my community. The Agers started their meat enterprise on Jamie's family land in 2000, inspired by Joel Salatin and others in the alternative agriculture movement, and began promoting grass-fed beef. Now you can find their meat at Whole Foods.

In Asheville, HNG is probably the best-known effort to scale up sustainably raised meat. "I was young when we started, and a different brand of idealistic than I am now," Jamie says. Since they started HNG, the Agers have taken many steps to market their grass-fed beef, pastured pork, lamb and poultry, including experimental production models and, in recent years, contracting with other growers to produce animals for their brand. These efforts to scale their local products to meet their customers' demands have been extremely stressful, exciting and full of opportunity as well as trial. The Agers goal is to create a brand that allows farmers to make money in a production paradigm that promotes high welfare, environmentally healthy livestock production systems. "I am still idealistic, to a degree," Jamie says. "I believe we have to change things in our food system."

The difference in the idealism lies in the knowledge Jamie and Amy have gained about the complexities of agriculture, and the flexibility and risk needed to face them head on. Not everyone can do it. "I've learned what an

ceived no subsidies and produces a small volume of pork under a completely different production system. These days, a pork chop is not a pork chop is not a pork chop. These are different systems, different products, different markets, different standards. Different prices.

Not everyone can pay more, *right now*, and not every farmer can happily embrace pastured, poison-free animals. I am well aware of this perplexing issue, and the fact that what we face is a giant, stinking problem. We may have the intention to create new systems, but the inability to do so. Many of us are stuck. Farmers can't pay more; most of the hungry, well-intentioned shoppers cannot pay more either; and so we

incredible amount of money, time and emotion it requires to move the needle on good meat," Jamie shares. This has bred in him a peculiar moderation, and an almost insatiable internal questioning about what to do and whether these systems are scalable. For Jamie and Amy, it's worth it to keep trying. They have seen positive change, and their buyers certainly thank them for it. "Every community, every farm and every market is going to look different, which is another thing that makes it hard," Jamie added. "At the end of the day, I'm a farmer," he says, "and I'm about farmers making money, and staying farmers. If the chance for a young farmer is to invest in a corporation, and put up a chicken house, so be it. He's enterprising in agriculture, and that is hard enough as it is."

I'd like to see us developing systems that are profitable for farmers, but better for the chicken and the diner as well. If it was easier for young farmers to enterprise in agriculture that was more sustainable for whole systems than that corporate chicken house, that would be ideal. In pursuit of this wish, I charge food citizens to regard the effort of farming more politely, and consider enterprising in good agriculture, farm education and research into sustainable agriculture. You can do this by buying local food, and you can take it even further by investing in local food and a local farm. It is not enough for us to ask our stewards of the land to shoulder so much of the risk of forging our new systems. After all, come dinnertime, farmers have to choose what food to buy, just like you do, after spending their long day choosing how much to fund better agriculture. It's a double whammy. In this way, local farms, butcher shops, bakeries, feed mills and other middle-system business owners are indeed our bravest pioneers in the journey toward better food. ◆

go, around and around, asking more of each other, blaming each other and begging forgiveness from each other, our refrigerators and frying pans all the while full of filth. This is a ghastly problem, but don't stop reading. I believe enough of us can pay more and enough of us are industrious—and to some of us, both apply. Luckily, eating animals is most rewarding to the industrious soul. And I believe that any person, farmer or not, regardless his resources or intent, could take something from this book to build better first- and middle-meat economies.

Cooking Differently

I remember speaking with a friend of mine who had opened two natural food stores in the foodie town of Asheville, NC, and asking him what department in the stores had the highest sales. His answer? Prepared foods. Of course, I thought; in our nation of convenience, in our culture of busy people seeking quick comfort. But if we're seeking truly good and honest food, we know we must cook more. We must be thriftier. We must learn to depend on ourselves again.

We not only need to cook more often, but we need to eat everything we're provided. The whole plant. The whole animal. Our selective use of food resources in America is so appalling that I lack an adequate adjective for it. And the ripple effects are endless, in the economy of the home, in our collective health problems, in our growing hunger problems, in our frenzied food production (more, more, more) and in our food waste management. A new approach to honest eating requires that we change this trend. Restaurants shall face this challenge, too. If there is only one hanger steak per beef carcass, it should not be a regular on the menu. Let's have better training in whole-animal butchery, so that we can feel comfortable seeing something like *herbed broccoli raab + date & sassafras sabayon + lamb* on a menu, and not need to know what the cut is, because the cut doesn't matter. The cut is whatever the chef needed to cook to make best use of the lamb's carcass. The fact of the matter is that poor land management and lack of attention to animal welfare has been bred in part by gross, disproportionate demand for rare muscle commodity. The industry and business of the external economy are built to respond in a language of brute efficiency, not complex consciousness. It takes people to manage business consciously, all across the supply chain.

In your home kitchen, learning to deal with a whole animal, or larger cuts of meat than retail-size ones, will necessitate ingenuity. You'll end up with rich stocks and meat that is best used for seasoning other foods. As you explore new ways to cook familiar cuts and adventure into unfamiliar creations, your mind will begin to unfold with ideas about pairing the meat with fresh vegetables, what fruit you may add to your porchetta or what herbs to try in the next sausage. You'll also find that meat is not always the main focus. Yes, you'll have some meat-centric, traditional American meals with those hunky ribeyes and roasts, but sometimes you'll get as much if not more nutrition and satisfaction if you let meat take back stage, using it to flavor, nuance and support the other food groups. If you can change your mindset about *how* to cook meat and *what* a meal looks like, you'll make excellent use of the whole animal.

Hopefully, in this journey, you'll begin to seek food that is more fresh and whole all around, and discover deliciousness without a lot of fuss. I have found that the confidence I have in the food at home and the joy I experience in the kitchen far outstrips the uncertainties of many experiences dining on the go. I have also found that it is easier to help people make full use of animal trimmings than it is to assist them in not wasting vegetable trimmings. Our collective allowance for vegetable waste is as appalling as our skepticism of animal offal.

Taking it further still, your ability to work with animal foods more competently will also change your desires about working with animal foods, which is a winning situation all around. For example, the muscles in a beef carcass you may not be aware of are now generally being processed into ground beef, a product we can all understand and afford. But if you learn to use the animal differently, you may find a way to eat those muscles differently, pay the same average price per pound of beef, but endure one less repetitive meal. And your farmer may have a greater chance of profiting off of one animal. This is just one example of how your cooking bone is connected to your buying bone, which is connected to the way your community looks, feels and functions.

Lastly, I want to talk about time, often cited as a reason we don't cook or preserve or really even taste anything we eat. I teach classes every so often about cooking from scratch and whole animal utilization. To do it, truly, is to save time and money. The meat trimmings or the braise liquid or the cut herbs from tonight may become a third of tomorrow's meal. Or you might throw them all into a jar with nutmeg and brine and use them in a soup next month. Although it will not come naturally to everyone to think like a chef, saving almost everything, your brain working as you go to determine what will happen with each remnant, it can be learned. Further, it can be enjoyed. This is the adventure of cooking. As your habits of ingenuity develop, you will discover creativity that you never knew. And as you expand your good food horizons, you will see that most good cooking can either be done ahead of time or done quickly.

I can't tell you how many times people have come over to visit and been astonished at "how quickly you got such a fresh meal on the table." To be astonished at the ease and simplicity of something like roasted pork tenderloin rubbed briefly with balsamic and herbs, flash-cooked brussels sprouts with pecans and buttered sweet potatoes, shows me that people have either had a) too much wine or b) are accustomed to spending only five minutes on dinner. I couldn't disagree more with the thinking that our food should be our last priority as we schedule our day or budget our resources. No matter where you go, food is the common denominator. And people's food is either killing them slowly or bringing them great joy, flavor and experience.

My good friends and close colleagues will attest that I have been guilty of a cynicism that often finds me saying things like, "The only thing that will cause us to realize good food again in this country is peak oil." It would be embarrassing to admit how much time I have spent pondering how big the natural or resource cataclysm will have to be to shake us into our senses. Even still, there is a tiny voice saying, "What if?" What if we could realize deliciousness, and honest living, out of our love for food and our desire for the journey and the experience it provides, rather than out of desperation? What if everyone just *knew* that the pursuit of better eating would make us better, happier people?

Cooking, and eating in general, should be one of the best things about our everyday existence. If it is a truly just a chore, a necessity, then we have surely sold our souls.

Eating Different Things

I am not a meat-crazed woman. I do not wear bacon T-shirts (although I think many of them are amusing), insist on meat at every meal or argue that meat is essential to every person's diet. I largely believe that each one of us is the proper authority on our own best nutrition. I detest dietary dogma, am extremely suspicious of mass nutritional trends, roll my eyes at the demonization or lionization of individual compounds or food groups and tend to laugh at diets that have names. I argue that the more diverse a diet, the better, as long as it is based on real, whole foods. Many people are surprised to discover that I can offer a class on vegan cooking.

I very much enjoyed Michael Pollan's book *In Defense of Food*, particularly because of his exploration of what, for years, I've been calling "the food Gestalt"—the assertion that food is more than the sum of its constituent nutrients, diets are more than the sum of their constituent foods and our health is more than the sum of our dietary parameters. In my classes, I talk constantly about the importance of diversity, and often struggle to succinctly cover all the empirical evidence that seems to support a diverse diet.

Let's look at nature, which most people can agree tends to *work*, even if it is beyond our ability to understand. Everything in nature is connected, and no one thing supports or destroys the whole. In fact, I charge you to close your eyes and imagine something that exists in isolation. You won't be able to do it. The entire world is comprised of wholes within wholes, all hitched together in an infinite feedback loop of diversity and synergy.

There are many classic examples throughout ecology that demonstrate how tampering with one element in a system will have ripple effects on the whole. We have seen this, for example, when one species is removed from an aquatic ecosystem, or a foreign species is introduced to a system. These sudden presences or absences change the balance, and like dominoes, a chain of cause-and-effect is unleashed as the system attempts to re-adjust and survive.

Now think about all the factors that drive your decision-making about your diet. A few of them might be taste, calorie intake, resource consumption or allergic tendencies. There are countless others, and more often than not we find ourselves compartmentalizing our meals because of them, rather than zooming out and thinking holistically. You might hear arguments alleging that your consumption of a food requires xx trillion gallons of water annually, so therefore you should eat something else instead, because water usage will decrease. And as you home in on water, or methane or gluten or antioxidants, you forget that you have slowly lost sight of soil, or omega-3 fatty acids, or God forbid, taste, until something happens to pull your focus that way once again.

How exhausting, to pedal and pedal a bike without the ability to step back and check whether or not it has wheels. I'd like to help you zoom out, seeking the viewpoint that gives you a sense of each whole, or system of wholes, and how they connect.

For example, instead of looking strictly at water usage per pound of beef produced, consider instead the water cycle: the system by which water is used, converted and recycled across the planet. For while it is true that water resources are being depleted due to overuse, and that most methods of producing beef waste a lot of water, this does not automatically make it horrible to raise beef, or use water in general. Consider instead the way that it is used, and how it may cycle back into the system to be used by all beings once again. Or how water as a resource can be affected positively along the way. For example, the act of animals defecating into pasture fertilizes the soil and provides more organic matter on the ground, which then increases the soil's capacity to retain water. As a result, water stays in the system and is made available to plants and other life forms. The animal, through many acts, is essential to the system. In this example, its poop alone is invaluable.

The automatic defensive argument is, of course, going to be that most cattle are not raised on pasture, so we are not seeing this manure benefit to the water cycle; instead, the beef industry is just wasting a lot of water. Too true. The equation is out of proportion. The challenge of holistic thought, and practice, is to recognize and reorganize the opportunities, and seek natural balance. Read: We cannot rid the planet of cattle. But we can change our systems to support planetary cycles and natural resource exchange. Poop of all kinds is a resource. We don't even begin to use it as such. This method of thinking can be extended to the other resource cycles essential to all life: mineral cycles, chiefly of carbon and nitrogen, and energy cycles. Holistic thought can happen on a small scale, to inform soil management on a fourteen-acre farm, and on a large scale, to inform waste management on an industrial feedlot.

It can happen in your home kitchen, too, and in your stomach. We ourselves are products of nature, and our bodies are complex systems of wholes. The very building blocks of our tissues, our cells, are complex systems of synergistic organelles. Beyond this fabric, we depend on other life forms incorporating their own systems into ours. Organic orchardist and author Michael Phillips illustrated this wonderfully in a recent visit to my hometown, when he said, "If you see me standing before you as one individual, one organism, you are wrong. I am standing before you as a community. And my eyeballs, teeth and intestines are all coated with trillions of microorganisms, without whom I would be a dead man."

It is so true. Your body and mine are all communities, wherein many individual beings live, all conspiring together to survive. Your gut alone is home to more than a hundred trillion microorganisms, a population that was mostly established by your diet and environment by the time you turned three. Without healthy internal bacteria, we would not be able to synthesize energy, extract minerals and vitamins from our food, digest properly or maintain immunity. We require a healthy, diverse "gut flora" to maintain a healthy body. Further, we are beginning to understand that our diets support or destroy the health and diversity of our gut flora, affecting our ability to avoid chronic diseases and allergies, build immunity to common illnesses and derive maximum nutritional benefit from our foods. Freeze-dried stool from healthy individuals is even being used to inoculate the digestive tracts of sick people and heal illnesses, and it is working.

This is just a slice of the thinking that translates directly into an argument for a diverse diet. Of course! If diversity allows a forest or water ecosystem to thrive, it will also allow our bodies to thrive. And further, we must seek our diverse foodstuff from the type of farms that are using holistic thinking in their management, supporting diverse and synergistic feedback in their energy, water and mineral cycles to grow our food.

I absolutely believe that abundance is possible, and that we can feed ourselves by emulating biodiverse ecosystems in our farming, mimicking natural processes as much as possible. This is the argument for regenerative, ecological agriculture, the opposite of the highly specialized monocultures that dominate our agriculture today. Agriculture that restores and respects the earth seeks to be as diverse as the systems the planet would itself create, and includes plants as well as animals, because each contribute in unique ways to the resilience of the whole. Attempting to manage one without the other is not realistic, and not correct.

As an example, I want to mention the work of Zimbabwean ecologist Allan Savory. After decades of research into ecological restoration, namely reversing desertification (the process by which fertile land is reduced to desert due to overuse or misuse), Savory has found that high-density rotational grazing of livestock is the only way to restore much of the earth's land. It works because the large herding animals mimic the wild herds with which grassland ecosystems evolved. A huge protestor of animal agriculture at the outset of his career, he has seen the effects of holistically managed animals on the restoration of soil, aquatic, grassland and prairie ecosystems on five continents. And with an estimated two-thirds of the world's land currently turning to desert (including the US's vast rangelands and land within our national parks), Savory asserts that holistically managed livestock herds are the only way to restore 95 percent of the world's land and feed our growing population.

We can use principles from the work of Savory and others pursuing integrated, holistic livestock management to inform our own production. While you probably don't possess a land base that can support twenty-five thousand sheep or cattle, and are likely not trying to feed Africa, you can manage animals to mimic nature, and do so in conjunction with vegetables, fruits and herbs, so that each system benefits the whole, maintains and regenerates the land base and feeds your family well. If you do not intend to raise your own animals, understanding the ecological and economic principles behind holistic farming will still assist you in making responsible buying decisions, ones that not only make sense for your health but also contribute to a more sensible and sustainable food system.

Lastly, a new approach to honest eating will simply require more diversity in your kitchen. Anything else would be costly, wasteful and disrespectful. Just as throwing out the celery leaves or the onion's succulent, green top is silly, throwing away perfectly edible parts of the pig just because you've not discovered how to eat them is uncalled for.

Ethical meat is not a utopian farce. It is real and mouth-watering, and can be healthy too. Considering the limitations and opportunities we face in search of good food, I see two paths in service of getting truly good meat on your plate. This book should serve anyone on either path, or a combination of the two:

1. Ethical meat will require a more robust first economy, wherein more people own the process for themselves, either through raising their own animals or buying whole animals and processing those animals at home.
2. Ethical meat will require us to cooperate on a deeper level with farmers and community butcher shops to source our meat, giving

the farmer and the customer more power to choose and boosting the middle economy.

You'll find, in this quest for ethical meat, that you will be able to improve your diet, impress your friends, unlock culinary creativity you didn't know you possessed and save money, all while eating extremely delicious food.

Regardless of the approach you take, you will need to skill up in either animal production, processing, butchering, cooking or preserving meat. Or all of the above. Let's get started, shall we?

The True Cost of Organic, GMO-Free Pork

This simple table compares total costs associated with raising a pig based on feed type. In the second column, you see costs associated with conventionally grown GMO feed. In the third column, you see costs of conventionally grown, GMO-free feed (while the seed is not genetically modified, the crop is grown with synthetic chemical herbicides and pesticides, and is not required to be managed according to organic standards). The fourth column lists costs associated with certified organic feed, which by law does not contain genetically modified seed and is grown with adherence to standards for environmental and consumer health.

These numbers are based on the enterprise budget of a pork and poultry farm near my home in Asheville, NC. My friends Graham and Wendy Brugh, of Dry Ridge Farm, raise about 150 hogs a year, plus about 1,500

	Conventional	Non-GMO	Organic & Non-GMO	Notes
Feed	$200	$350	$450	3.5 lb. feed produces 1 lb. meat
Non-Feed Costs	$18	$18	$18	Equipment, fencing, labor, etc.
Slaughter	$50	$50	$50	Kill & dress
Processing	$200	$200	$200	Cut and wrap
Total	$468	$618	$718	
Total per lb.	$2.84	$3.75	$4.35	Based on 165 lb. dressed weight
Sale Price per lb.	$4.70–7.10	$6.25–9.40	$7.25–10.88	Wholesale-retail range

chickens. They also sell lamb. These numbers allow them to sell weekly retail at farmers' markets and support some wholesale accounts with local butcher shops and grocers. Note that the price per pound of feed is based on the price their farm gets, since they buy their feed by the ton. For those raising pork on the home scale and thus buying feed by the bag, prices per pound will be slightly higher.

In the Notes column, on the first row, you'll notice a mention of feed conversion ratio. This is an important number in livestock production that refers to the amount of feed, in pounds, that the animal requires to gain one pound of meat. The ideal feed conversion ratio for pork is 3.5:1, meaning it will take you 3.5 pounds of feed to produce 1 pound of meat. Note that this ratio can vary greatly, depending on breed, feed quality and herd health. Using ideal ratios in this example, it will take the farmer about 962.5 pounds of feed to take one animal to a finished weight of 275 pounds.

Non-feed costs include equipment for fencing and watering, labor for moving and caretaking and other animal needs. Again, keep in mind that as volume goes up, price goes down. The more animals there are on the ground, the lower the non-feed cost per animal.

Slaughter costs include killing and dressing, which is removal of the animal's innards. Some organs can be requested, but depending on regulations in your area, or the rules of the processing house, you may not be able to keep organs and blood. The example here omits organ and blood weight, assuming they are lost in the process. This is common in my area. If the slaughterhouse can sell them, perhaps they do, but the farmer can rarely make use of them, as regulations either prevent her from getting them back in the first place, or limit her ability to do any further processing (of blood sausages, salamis and pâtés) without having her own inspected facility.

Thus the final row, processing costs. This is the cost to "cut and wrap" or turn the carcass into chops, sausages and other retail units. This happens at the slaughterhouse, by regulation in my state, or at a facility permitted to further process meat, such as a butcher shop or grocery. The cut-and-wrap cost is one of the highest costs in small-scale livestock production.

In the final rows, you see conclusions regarding total cost per carcass, and then total cost per pound of meat. Note that the cost per pound is based on *dressed weight*, the weight of saleable meat once the animal is killed, bled and dressed. This is different from the earlier multiplier, the *finished weight*, which is the production goal the farmer seeks to reach before taking the animal to slaughter. For this example, I am figuring a dressed weight of 165 pounds, meaning the farmer has lost 40 percent of the carcass during processing. The losses include blood, some bone, all the innards and, in many cases, the head. If the animal is shot with a gun rather than stunned and bled, the head cannot be used, and is added to the waste can.

My hope is that consumers of ethical meat can look at this and begin to see the economic conundrum in plainer view. If the farmer is to make a living off of production, he or she must charge 40 percent more than the cost per pound to wholesale customers, and up to 60 percent more than cost to retail customers. This is the arithmetic used to figure the average range in sale price per pound.

If you're still with me, consider that not all parts of the carcass are created equal. Per carcass, the farmer can expect to produce 25 lb. loin, 35 lb. shoulder, 25 lb. ham, 25 lb. sausage, 12 lb. shank, 20 lb. belly, 5 lb. ribs, 10 lb. fat, plus feet and maybe head. Some parts of the carcass are in higher demand than others—for example, customers prefer loin chops over leg shanks, so the farmers cannot apply the same price premium for the entire carcass. (You would not pay $7/lb. for feet or back fat). So the markup is higher for meats that are coveted, and lower for lesser-used parts. This is all designed to permit the farmer an average price per pound on the total animal, which ensures his or her success as a businessperson.

You can easily see how much of a premium the higher quality feed produces, and how this translates into price. For the farmer in business, options are limited. As any smart businessperson knows, if you cannot turn the cost increase over to the customer, you must limit costs in production. Looking at our table above, the highest costs are in feed and processing, both enterprises that are mostly out of the farmer's control. Even limiting the non-feed costs will not aid much in the end result, and increasing the number of animals may help some but requires sufficient land (which the farmer may not have).

He probably does not have the land or equipment to grow his own organic feed, nor does he have the capital or desire to build a feed mill (for grinding and mixing feed rations) or develop distribution for the feed once it is grown and milled. The feed mills that do exist are having a hard time selling premium feed in enough quantity to justify a better price, as their own costs in buying, grinding, mixing and delivering grain are also delivering tight margins. You see, feed mills are also small businesses, also trying to strategize about economy and quality so they can survive.

On the processing side, the farmer is up against limitations in regulation and customer expectation. If law dictates that she cannot keep or use the innards, and if the market does not demand blood sausage anyway, she loses a percentage of product. If slaughter practice and regulation does not permit her to keep the animal's head, and the customer will not buy it either, she loses a percentage of product. If she does not have the knowledge or infrastructure to cut and wrap her own animal, she is forced into paying a premium for cut-and-wrap services, which further gouges her profits.

Where is the answer? There are perhaps a few. Feed cooperatives and tighter communities of farmers and associated feed mills would increase the volume of organic grains sold, potentially lowering prices across the board. Meat cooperatives and tighter communities of farmers and associated consumers could increase the volume of whole animals sold or cut-and-ready meat produced, lowering prices across the board. But the answer that is most readily in reach is you, the enlightened consumer. If you will buy a whole pig or half a pig and butcher it yourself, you will help the farmer eliminate processing expenses. If you will eat delicious pâtés and headcheese, we can make better use of the valuable animal. ◆

General Notes on Raising, Cooking and Eating Animals

Slaughter

Ideally, we would complete this most difficult part of animal production ourselves. This book does not have the capacity to offer instruction on slaughter for large animals, but I recommend books and materials in the Resources section that do.

Temple Grandin, renowned livestock handling expert, has said, "I believe that the place where an animal dies is a scared one. The ritual could be something very simple, such as a moment of silence. No words, just one pure moment of silence."

If you can't do the slaughter yourself, take good care to find a facility that will do it in the least cruel way possible. This entails the company of the person who raised the animal, a calm transport to the slaughter facility, and calm entry into a holding pen. The best facilities use a stun gun to render the animal unconscious immediately. These devices fire a blank into the animal's head, between its eyes, or apply blunt force at the base of the neck (in pigs). Once the animal is stunned it is hung and its throat cut, so that blood drains out before the innards are removed, a process known as "dressing."

After it is dressed, the animal is skinned or de-haired and halved. At most USDA- and state-inspected facilities, sides of beef are then aged for at least fourteen days in cold storage before being wrapped and shipped or further cut.

Because cattle are such large animals, I recommend that homesteaders who do not slaughter on their property have the animal cut into quarters, or even further (discussed in the next chapter). Assess your workspace and equipment before deciding how much you want the processor to do for you. If you have a standard-sized, six-foot worktable, you will likely need primals, as the larger quarters of beef can average 150 pounds or more. For pork, you can choose to have the processor break each half into three pieces, or you can work from intact halves. Lamb, goat, poultry species and rabbit are all manageable in as whole a form as you can finagle from the processor.

Notes on Cooking and Eating Muscle

Regardless of species, we must recognize the impact an animal's life has on its muscle and on the eating experience. What the animal eats, how it moves, how old it is, how it is handled at slaughter, the composition of its parents and the breed from which is arises are just a few of the many factors that affect muscle quality. No two animals are the same, ever. Nature is endlessly various, which is why farming, butchery, charcuterie and cooking are all art forms that seek to work with the fascinating and dynamic materials nature provides.

In general, the muscles that the animal uses more frequently tend to be leaner and have more blood flow to them, and thus more flavor. In cooking, this translates to tougher but more succulent cuts that should be braised, slow roasted, stewed or smoked. The muscles that the animal uses less frequently will be fattier, tenderer and less flavorful. These muscles should be cooked quickly using high heat, as in broiling, grilling or frying. Similarly, younger animals will have tenderer, paler muscles, while older animals will have moved more, put on more fat and gained more flavor.

One thing you absolutely cannot do without is a meat thermometer. If you plan to eat meat and do it right, this is an indispensible piece of kitchen equipment. No compromises. Everyone at my shop used to ask, "How long do you cook that for?" when tasting the meat we had on our menu. Our serious-joke answer was always, "Until it's right." Meaning, there is no time prescription, usually. It depends on the size of the muscle, the smoker or oven temperature and the preparation. The internal temperature of the meat is the best indicator of done-ness: 145°F for pork, 155°–160° for poultry. Beef and lamb is not as good when cooked that completely, but if you are going for well done, take it to 140°. I prefer to take my beef to 120° or 130°, and then remove it from the heat. All meat will continue to cook after it comes off of the flame.

With most of the recipes in this book, I've chosen to focus on cuts that you're probably less familiar with, or more likely to have trouble with. I will touch on basic techniques for different parts of the carcass in the butchery sections, based on muscle origin, but the details will focus on the lesser-used parts.

For general guidelines on smoking and sausage making, refer to the charcuterie section first.

Disclaimers

On Farming: Production notes for each species are intended as overviews only. The logistics and science behind best management practices are much more complicated than this book can possibly contain. Out of respect for the animals themselves, and out of a personal loathing for literature that oversimplifies farming and responsible land management, I have compiled a Resources section that I hope you will use to further your education on production techniques. I believe the best farmers learn by doing, but I also believe there are solid scientific principles that should not be taken for granted in caring for animals, plants and the soil.

On Cooking: It is very hard to write recipes. I cook using my senses, and often do not record exactly what I have done. To put exact numbers on ingredients and try to perfect so many ratios has been a challenging process. I urge you to come at it organically, and use your intuition and sensory wisdom. The more you can do this, the better off you'll be in the kitchen, anyway.

Note to Reader: All recipes for fresh meat have ingredient measures by volume. This is the way most people cook. However, weighing ingredients is the most precise way to prepare food, and it becomes essential when you are curing products, or working with fine-tuned ratios in sausage making. As such, all sausage and cure ingredients are noted by weight.

Butchery Tools and Tips

If you're ready to try your hand at whole-animal butchery, you need to get a few things straight. Your tools are important—and dangerous. Diligent care and good technique are the two things that will keep them, and you, intact. Here are a few rules, which we'll see in action as we discuss butchery of each species:

- Cut across the grain of the meat, and keep muscles as whole and undamaged as possible.
- Leave as much fat on the muscle as you can muster. It will turn into deliciousness. You can always remove fat after cooking, if you desire.
- Use big knives to break and push, a boning knife otherwise.
- Never saw through meat. Saw only on bone.
- Keep your stance grounded, moving the meat to assist you, when possible. You don't always need to run around trying to get the best angle.
- Use sharp knives, always.
- Keep conditions sanitary. Make sure surfaces are disinfected at start, clean blood as you go, and always keep meat as cold as possible.

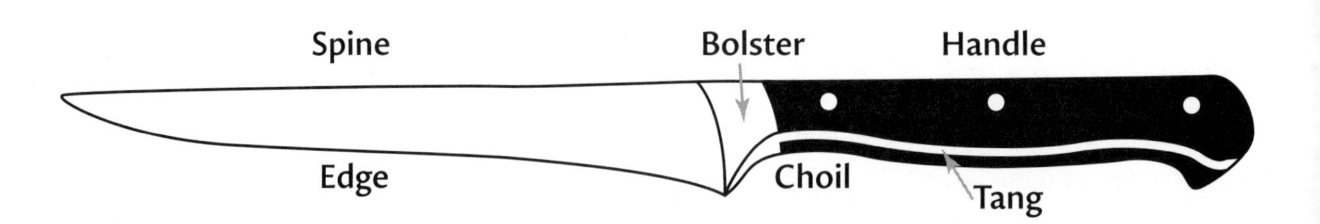

As you proceed in your practice, you'll develop preferences for knife grips, knife varieties and other equipment. I've listed my tools below for your reference.

Knife grips are worth mentioning. There are three main grips. While you may be more comfortable making variations on these three, note that over time, creative/renegade handling can lead to arthritis and bone spurs. Butchery is very hard on the hands. The three main grips:

- **Pinch:** your hand clasps around the base of the spine and the bolster, and the upper part of the handle. Used for slicing and boning.
- **Thumb:** Your thumb is braced against the back of the knife's handle, while your fingers curl around the front of the handle. Used for boning and pushing through meat.
- **Dagger:** Your fingers are curled around the back of the handle, your thumb curled around the front. Used for big breaks and pulling through meat.

You'll use your *boning knife* most. I use a Forschner-Victorinox semi-flexible 6-inch boning knife most of the time, but for trimming and work on retail cuts, I also have a stiffer Messermeister (German) forged knife. Semi-flexible, stamped knives are ideal starter tools.

You'll also need a larger knife, such as a *butcher's knife* or *cimeter* (pronounced like "simitur"). I use these interchangeably, so you need only invest in one or the other to get started. I had my cimeter hand-forged by my friend G. Kearney at Kearney Knifeworks near Winston-Salem, NC. Forschner-Victorinox makes good breaking knives to get you started.

If you want to keep your knives for a while, *a honing steel* will help you along the way. You can buy these at kitchen stores, or order online. Since your cimeter is likely to be at least 12 inches long, you'll need a 10- to 12-inch steel; anything shorter will be a waste of your money. Figure out a plan for *sharpening your knives* as well. You can send them off to have them sharpened, find a local butcher shop that will do it for you, or sharpen them yourself. Grinders and electric sharpeners are expensive, and it is easy to buy something that will wreck your blade. I use oilstones to sharpen by hand.

A *bone saw* will be your go-to for getting through ribs, spines, arms and legs. A 25-inch meat saw will be versatile enough for any species you choose to tackle. Beginners, or folks interested in smaller animals, may choose to look for a more manageable saw, something along the lines of 19 inches. Butcher and Packer (butcher-packer.com) is a good source for these, as well as replacement blades. In fact, this company is a great start for many of your equipment needs.

You'll need a *cleaver* for getting through rib and spine in some applications. Dexter-Russell makes the best cleaver. You'll need a *rubber mallet* to pair it with, and you can pick these gems up for under five bucks at the local hardware.

A *bone scraper* or bone duster is an optional tool for removing bone fragments from muscles after sawing. If you choose not to invest at first, you can achieve this with a damp rag. *Butcher's twine* is another handy item as you go.

Meat hooks are helpful as well. You can get them with handles, at which point it is proper to call them *boning hooks,* and use them as additional grip when you are butchering large portions or working "on the rail" (butchering a hanging carcass). We'll discuss butchering "on the bench" in this book, assuming that your best option is to keep the carcass on a work table in front of you. Other meat hooks you might choose to acquire are *s-hooks, ham hooks,* and *rail hooks,* for hanging portions in the cooler or hams for curing. There are also *bacon hangers, belly spreaders,* and *gambrels,* which are often used for hanging smaller portions or carcasses, curing bacon and salami, and making other cured items.

Your *worktable* is an important consideration. Since you're planning to work with larger portions, you need space. Half of a pig can sometimes be eight feet long, and a quarter of beef can easily be four feet by eight feet, if not bigger. What you can accommodate weight and space wise will determine how much you need your meat processor to break the animal before you receive it. At bare minimum, set up a six- to eight-foot table on level, sturdy flooring. A table top that is about six inches above your wrist when your arm is held straight down and relaxed at your side is a good working height for butchery. Some may prefer it slightly lower, to help them bear down on larger portions of meat. In general, it is not a great idea to have your table against a wall. Free space around the table is best, so you can move around larger portions and big bones and considerable girth won't keep knocking the wall and keeping you from moving the meat around freely.

Cutting boards will go atop your worktable, and can be sourced of polypropylene or wood. If you choose wood (it's the best), look for end-grain maple construction. Your knives will thank you for it. Boos is the leading manufacturer of maple butcher blocks. If you have a limited budget, look for commercial-grade, large poly cutting boards, at least a half-inch thick. Make sure your boards won't be slipping around while you work, by placing damp rags under them or wrapping your table's top with plastic wrap. The latter is a real lifesaver with cleanup to boot.

Wear close-toed shoes and sturdy clothing. Knives slip and get knocked off of tables, and you will make mistakes. The question is not whether you will cut yourself, but rather when; don't take your toes or fingers for granted. You can opt to use a *cutting glove* on the hand that you usually use to hold meat steady. These are made of chain mail, Kevlar or other high-density synthetic materials.

Put some thought into *storage* supplies for your meat once it is processed. Plastic wrap and butcher paper will only get you so far. Consider investing in a small vacuum sealer, such as a FoodSaver, especially if you intend to freeze a lot of meat. Vacuum sealing can also be a good way to set cures, if you intend to venture into preservation.

You'll definitely want a *meat grinder* for making sausages. There are several tabletop models available. I like LEM's products. For this book, I wanted everything to be small-scale and appropriate for the homestead, so all these recipes were tested using a Chef's Choice stainless steel grinder attachment for a standard KitchenAid mixer. I found that this worked pretty well for five- to ten-pound batches of sausage.

Get a *sausage stuffer* that you can use by yourself. I use an Omcan vertical stuffer, and I hear water stuffers are good for working solo. I advise against the stuffer attachments for mixers, as they require you to shove the meat into the stuffer as you work, which only invites unwanted air into the mixture. Vertical stuffers allow you to put all the meat into the hopper before you start and use a press to keep the mixture tight as it moves through the machine.

Casings are either natural (from the intestines and organs of beef, pork and sheep) or fibrous (made from collagen or other materials), and you'll use different types for different applications. I strive to use natural casings as much as possible. I use natural hog casings for fresh sausages, sheep casings for breakfast-type links, beef middles for most salamis, and beef bungs and hog bladders for some of my whole-muscle ferments. I get all my casings for home production from butcher-packer.com

You may also find yourself looking at *netting and bags*, for containing some of the larger sausages and cured items in the charcuterie cabinet.

A *spice grinder* or old coffee grinder is essential for sausage making and general mad scientist fun.

A *food processor* is indispensible in any kitchen, in my view, but especially if you're interested in making emulsified sausages or pâtés. You may also want to invest in an *immersion blender*. I use both all the time.

Loaf pans or *terrine molds* for making headcheese, terrines, pâtés, and rillettes.

Cheesecloth for bouquet garni and for wrapping meat before hanging.

pH strips (cheaper) or a *pH meter* (better) for checking the status of your charcuterie projects. Hanna is the generally recommended manufacturer for pH meters with calibration capabilities. They run upward of $300, so for starters you can look into the colored test strips sold at hardware stores. These are also used for testing the water in home swimming pools, so anywhere you can get pool chemicals you can usually find pH test strips.

A digital g and oz. scale is definitely needed for getting proper ratios in sausage making, and for getting start and finish weights on cured meats.

Butchery Tools Checklist

- boning knife
- cimeter or butcher's knife
- bone saw
- cleaver
- rubber mallet
- bone duster
- butcher's twine
- hooks (boning hooks and s-hooks)
- sturdy table, at working height
- cutting boards
- cutting gloves
- vacuum sealer and vacuum bags
- honing steel, for keeping your knives up
- plan for sharpening your knives
- meat grinder
- sausage stuffer
- bowls for mixing
- spice grinder
- cheesecloth
- food processor
- casings, netting and bags
- loaf pans or terrine molds
- pH meter
- digital scale

CHAPTER 1

Beef

I've been working on this steer forever. He's the last one left in Rocco's herd, and he won't move the way I want him to. He's Charolais. Beautiful, white with gray eyelids and a gray nose, hooves. I've got my six-year-old on my back, saying "Go on, cow!" not really loud enough for the animal to actually care. He's looking me right in the eyes. I can tell he's lonely, hanging out with all the sheep, so he does what he wants. I can relate. Sometimes he lounges with the donkeys, but he doesn't feel quite right about that either. We're under a small grouping of trees, and he's alternating between eating and staring at me. I am alternating between backing away and advancing, in an effort to move him between paddocks, but he only sidesteps me and sniffs.

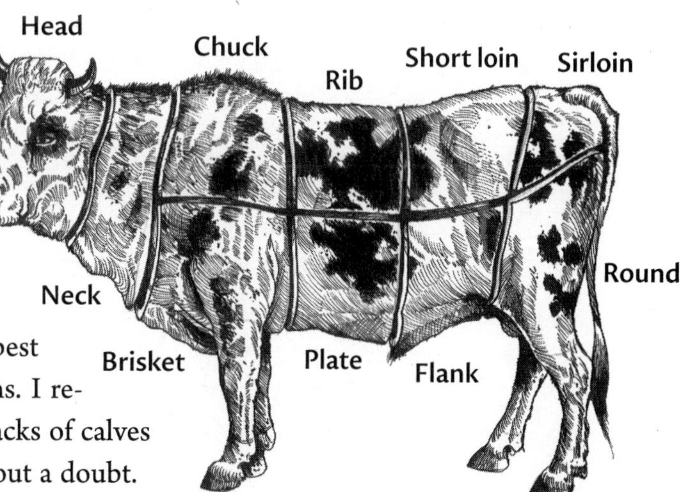

Looking into his face, I'm flooded with memories of my own cattle herd, which we slowly sold off as our farm began to fold. I remember pulling calves in the woods while I myself was pregnant. I remember walking the cattle paths in the pasture in early morning, thinking the cows had the best view of the mountains, and the seasons. I remember running my hands over the backs of calves just born. Cattle will change you, without a doubt.

I think the first time I stood in a pasture with Tex, our first heifer, I felt a mixture of curiosity and terror at the sheer size and sight of her. Mostly, though, I remember the uncanny peace she taught me over the years. I remember the wonder that filled me when her first calf sprang up on wobbly legs in the grass, while my own firstborn remained heavy-headed and dependent. In many ways, cattle convinced me of the grave inferiority of humans in the huge and sweeping natural fabric. Spend enough time in a field with cattle and you will silence. You will ease up.

I get the same feeling from this stubborn beef, here. In fact, if there wasn't so much thistle and black nightshade underfoot, and if my boot hadn't come apart at the toe a second ago, I'd have half a mind to lie down right here in the pasture and stay awhile. But my son is bleating softly, still on my back, about how it is getting hot. He's thirsty. "Just look at him, momma. He's not coming." Yep. The kid's right. We stay a stitch longer, just looking him in the eye. "Look at him, Cash," I say to my boy. I hope he sees the resolve in this animal's face. I hope he sees the intent toward a long, silent day in the sun, with fresh grass in the nose and no one calling. I know I do, and I'm not gonna hurry the day now. Nothing much is too pressing.

A few more advances, a few more sniffs. We bid the steer a good morning, leaving the gate open for him to change paddocks at his leisure, and head back to the barn.

Raising Beef

Most of the beef consumed in our country is raised in feedlots or animal feeding operations (AFOs), meaning animals are managed in confinement without access to grass. The classification "Concentrated Animal Feeding Operation," or CAFO, applies to about 15 percent of all AFOs, and is determined by the stocking density of the animals as well as the amount of waste pollutant produced by the operation. By definition, an operation confining at least 1,000 beef cattle for at least 45 days in a 12-month period in an area where they do not have access to grass is big enough to be considered a "Large CAFO." However, it is not unusual to find CAFOs that house closer to a million animals.

The typical American beef cow starts its life on a ranch, usually feeding on its mother's milk for 6 to 10 months while roaming on pasture. At weaning, the calf will be sold at auction, where it then proceeds to the feedlot. Here the animal typically consumes feed ration from troughs until it reaches 18 to 22 months or a finished weight of 1,200 to 1,400 pounds. Then it is sent to the processing facility, where it is slaughtered and inspected and its meat packed for sale.

But there's more than one way to raise beef. You've probably heard production claims like "no added hormones," "no added antibiotics," "all natural," "pasture-raised" and "grass-fed." (See the sidebar Label Claims for a breakdown of common claims on meat products, p. 38.) Chances are, if you're looking to raise your own beef, you're looking to take advantage of the more natural tendencies of cows, which favor a grass-based or grass-only diet. We'll discuss both. If you're not looking to raise your own animals, pay careful attention to the label claims, and if you're asking me, seek the most local products possible, so that you can talk to the farmer about his production standards.

Breeds

As with any animal, consider your needs in order to determine beef breed. Hardcore homesteaders will likely consider dairy breeds, as these cows will provide ample milk and tasty meat to boot. Some of the best beef I've ever eaten came from dairy breeds raised as milking animals and eventually retired to the freezer. Dairy cows are large animals, in general, with big bones and lean frames, and some people will not like the taste of their meat, especially if you plan to feed strictly grass. If you are just getting started and are looking for a dual-purpose animal, consider the smaller of the dual-breed or dairy cows like Dexter or Jersey. These breeds average from 600 to 900 pounds, as opposed to the 1,200–1,500-pound Brown Swiss or Guernsey animals.

If you're interested in strictly beef on the homestead, consider Devon, Gelbvieh, Charolais, Limousin or Angus. The farm where I work part-time outside of Asheville is experimenting with an African breed called Mashona, and seeing promising results for grass-fed beef.

Label Claims

If you've sought ethical meat before, you've likely run across a battery of different label claims that can leave you feeling lost. Here's an overview of the most common claims you'll see when buying meat, and some information on how they stack up.

No Added Hormones: In most AFOs, animals are administered hormones mixed in their feed to speed and instruct their rate of growth. These hormones are either endogenous (meaning they are hormones humans and animals produce naturally anyway) or exogenous (synthetic hormones made to mimic natural ones). A claim of "No added hormones" means the producer has not introduced any hormones, either endogenous or exogenous, during the animal's life. The United States Department of Agriculture (USDA) prohibits the claim "Hormone-Free," so if you see this on a label, be aware that it is uneducated at best. All meat has hormones in it—it comes from a live animal with a functioning endocrine system, just like you.

Be aware also that there is no organization or entity that verifies "no added hormones" label claims, or keeps a record of claims and practices. It is the marketing agency or producer herself that make these claims. The USDA can, however, verify a "No added hormone" claim made by a producer, if the agency deems this to be necessary.

The "no added hormones" claim really applies to beef only, as the USDA has banned the use of added hormones in pork and poultry production. If you see a claim of "no added hormones" on a pork or poultry product, know that the producer has not made any extra efforts outside of the standard regulatory requirements for the raising of those animals. This is not to say that all pork and poultry labels claiming "no added hormones" are out to dupe you. Many producers are aware that consumers are ignorant to standard regulation, and as such seek to demonstrate the purity of their product at the point of sale, regardless of whether it is a special claim or a standard one.

No Antibiotics Administered: In most AFOs, animals are administered low-level or routine antibiotics in their feed ration to combat possible diseases, which can arise from their rapidly changing gut composition and the high density of animals and feces within the containment area, which exposes them to potential infection. The claim "No antibiotics administered" means that the producer did not provide any low-level or routine antibiotics during the course of the animal's life.

This claim is important, as the evolution of bacteria that are resistant to antibiotics is controversial. The ramifications of antibacterial resistance are widespread in both human and veterinary medicine. According to the Center for Disease Control and Prevention (CDC), upward of two million people are infected annually with bacterial diseases that are resistant to antibiotic treatment, and 23,000 people die as a direct result each year. Antibiotics sold to the animal agriculture industry outnumber antibiotics used to fight infections in humans by a factor of four.

The USDA has banned the claim "Antibiotic free," so if you see packaging with this claim it is neither legal nor verified. The USDA oversees verification of claims of "No antibiotics administered."

Grass-Fed: The claim "Grass-fed" means that the animal had a strictly grass diet, with no supplementation of grain. The American Grass-fed Association has the highest standards for production, so if you see their seal, you can be sure it is the real deal. The USDA also verifies this claim. And you will sometimes see the USDA's shield on a package, either with or without an American Grass-fed mark. Be aware, however, that with the USDA seal only, standards may be less stringent, as the claim may have been grandfathered in from before USDA established standards for production. Before 2007, the definition was vague and there were fewer corresponding standards for producers.

This claim is important if you prefer grass-fed meat, but applies to beef and lamb only; there is really no 100-percent grass-fed pork or poultry. The controversy surrounding grain supplementation focuses on whether grass-fed meat is of greater nutritional benefit to humans, and the concern over genetically modified grains.

All Natural: This claim has no meaning, really. The words "natural," and "naturally" are used all the time to trick consumers into buying products that have no special production standards applied. This is not to say that someone using the word "Natural" in his or her farm name or product labeling is not providing a healthy product. Talk to the producer or research practices to find out more. The point here is that "All natural" is not a regulated or verified label claim, and there is no standardized definition of what it means for human health, animal welfare or environmental stewardship.

Animal Welfare Approved is a non-profit organization. Their certification program audits farmers annually to determine best management practices for physical and psychological well-being of animals. Their standards cover humane breed selection, the absence of feed additives and the respectful treatment of animals. When I was farming, AWA was the best standard for humane and healthy treatment of animals. Even better, they certify only family farms or family farm co-ops.

All in all, AWA is a label claim you can feel extremely good about.

Pasture-Raised: "Pasture-raised" is not a verifiable or standardized label claim, but it is showing up more and more in the meat world. It refers to meat that is raised on pasture, with constant access to grass, but receives either a free choice grain supplement or a finishing ration of grain. The term "free choice" in feeding means that grain or supplement is available to the animal at all times, and the animal eats it as desired. A finishing ration is a grain feed given to the animal in the last part of its life (in beef, usually up to 120 days) to help bring the animal up to finished weight.

There can be a lot of smoke and mirrors with this claim. If you look at the life cycle of the typical American CAFO beef cow, you'll notice that the animal is most likely born on a small(er) cow/calf farm, where it roams on pasture and feeds from its mother's milk. Technically, if I ran a large CAFO and purchased weaned calves at auction from a small grass ranch, I could market my finished animals as "Pasture-raised." However, family farmers and producers with small, fully grass-based holdings are finding that this claim is a good conversation starter in terms of educating customers about how their meat is different from the status quo.

At some point, the term "Pasture-raised" will probably get more attention from regulatory bodies or concerned consumer groups, simply because it is not regulated. There are pros and cons to standardization, as always. Once we put rules behind words, we open up a whole can of worms in terms of profitability and practicality, and set the stage for big-marketing dog and pony shows, loopholes against consumer safety and farmer fairness and profitability.

For now, I'd urge you to give "Pasture-raised" some good attention. Ask questions and verify sources. In my opinion, meat from honest, upstanding farmers using pasture-based systems is some of the best meat you can buy. In many ways, this production model strikes the best balance in our current world as far as healthy, conscious meat *and* farmers who stand a chance at staying in business.

Free Range or **Free Roaming** means that animals have had access to the open air and can move around. It does not mean they ranged on grass, although access to the outdoors is the standard defined for meat poultry (not eggs). The USDA definition states that five minutes per day is enough to justify the claim, so it has little meaning. Other than with meat poultry, the term is not defined and has no standards behind it, so it is mostly a marketing claim. Again, talk to the farmer if you can, to determine how he or she defines "free range."

American Humane Association is a third-party certification of animal welfare practices; however, producers have to score only 85 percent to comply. Additionally, the standards allow for the potential overuse of antibiotics, since an entire group of animals can be treated if just one animal is sick. They also do not require open ranging for animals, outdoor access or production standards that allow animals to carry out natural tendencies. All in all, this is not a reliably consistent or meaningful claim to the honest meat eater.

Certified Humane Raised and Handled is a standard of the Humane Farm Animal Care Organization. The standards prohibit added hormones and antibiotics in production, and require animals to be raised in clean and healthy systems. The program includes protocol for environmental stewardship, and even has standards for slaughter facilities that are more stringent than the federal standards for humane slaughter. Third-party certification is required for producers and processors using this claim.

USDA Certified Organic is a highly respected label claim administered by the USDA's National Organic Program. Standards for livestock require that animals be fed grass from pastures managed organically or grain from certified organic sources. As such, meat stamped with the USDA organic seal is guaranteed GMO-free. The standards prohibit the addition of hormones and antibiotics, and producers and processors using the seal must be third-party certified.

However, there is no doubt that large, corporate farms have taken advantage of loopholes in the USDA Certified Organic standards that compromise animal welfare, particularly with respect to stocking density and access to grass and the outdoors for non-ruminant animals. It is also rare to see the USDA organic seal on meat coming from small-scale farms, as the certification process can be tedious and expensive. ◆

A few butchers I've met and worked with are really interested in the Devon-Angus cross.

As you adventure further into cattle production, you'll find die-hard breed loyalists, and you'll find folks who swear by their mutt herds. If you plan to grass-finish, there is a decent amount of science behind which breeds make the best grass-fed beef. I know I have had really excellent grass-finished beef as well as pretty horrible grass-finished beef. While genetics surely came into play, the quality of the forage and the age of the animal carry equal, if not greater, weight.

If you plan to supplement the animal's diet with some type of grain or protein, you may be less concerned with breed selection. A discussion of the role of genetics in beef production is fodder enough for its own entire book, so again, consider breeds purely according to your preference. As you go, you will develop your own opinions about the importance of genetics to taste. Beginners should base their initial breed considerations on the animal's size, temperament and ease in calving (if you intend to grow your herd).

Space and Water

Figure two acres per animal. This will provide enough space for rotational grazing, wherein you move the animal to fresh pasture to maximize the forage's growth as well as the animal's nutrition. Many graziers are stocking at a much higher rate, and moving frequently, to mimic grassland herds.

Water is one of the most important considerations when raising any animal. Responsible production requires fresh, clean water for all animals without depleting natural water resources. Do not let your animals drink from the creek. Their manure in the creek is a disturbance to the natural aquatic ecosystem, and can introduce harmful bacteria that will travel with the water. Fencing the animals out of nearby water bodies, but tapping into these water resources via piped irrigation or pumps, is a cost and effort that you will have to assume. See the Resources section for further reading on setting up watering systems and equipment to make watering and rotating more efficient.

Fencing

In my experience, cattle are not difficult to contain, *as long as they have grass to eat and water to drink.* There are several options for fencing. Before choosing a system, consider whether you will want to add other species to the same pasture, like sheep, and think about what fencing will be most adaptable.

Electric fencing is ideal, although many small-scale growers opt for barbed wire or split rail for just a couple of beef animals. Either way, source good wooden posts of quality, rot-resistant wood, like locust. You can also use metal T-posts. Note that if you do plan to add sheep (or the ultimate escape artists, goats), you will need to upgrade, as these fences often will not contain smaller ruminants. Also consider predator pressure. Are your animals in a pasture far from your watch? Predators are less of a concern with beef, but if you are breeding, coyotes can target younger animals.

In commercial, pasture-based systems, standard fencing for beef cattle is a high-tensile (permanent) perimeter fence and poly-wire (temporary/movable) cross-fencing for rotational grazing. A high-tensile fence is made with wooden posts and braces and heavier gauge wire under tension, which carries a current from a solar- or battery-powered charger. The poly-wire is polypropylene strand with a thin wire running through it to carry the current. Farmers hook the temporary poly-wire fences into the perimeter fence to create paddocks, or grazing areas within the pasture. The posts for these poly-wire fences are typically plastic, fiberglass or insulated metal step-in posts that can be easily removed and relocated. Polypropylene fencing also comes in nets, which is great for poultry and smaller ruminants.

A great system for starters would be a three-strand poly-fence with metal corner posts and fiberglass posts elsewhere. It will take some investment, but you can adapt it without too much additional cost as your system changes. You could also use poly-netting as opposed to three strands of wire. Premier 1 Supply is a great resource for fencing equipment and instructions on fence design. See the Resources section for additional reading and supplier information.

Feed and Minerals

What you feed your beef cattle, or any livestock, is one of the most daunting and important topics in food production. Necessarily so, as it is nearly universally agreed that the quality of food the animal eats is equal to, or of even greater importance than, the quality of the meat it provides. I'd argue that it's of greater importance. When you choose the source and quality of feed for your animals, your decision to eat meat starts to reach out and tap hundreds of other systems and beings, and becomes not just a relationship between those animals and you but a relationship with the whole world. Those who eat plants only are not exempt from this butterfly effect. Let us not be ignorant to how our need to eat, and our decisions regarding what we eat, both harm and nurture many, many living individuals.

We've seen that in the highly efficient, industrialized systems of beef production, diets high in grain are favored over diets based on grasses. This allows for less land area to be consumed in the production process, and guarantees faster weight gain. But research shows that consuming so much grain, as well as being exposed to harmful bacteria and subsequent medication, changes the animal's gut chemistry, affecting its health and well-being as well as the health of the meat it produces.

Just as you depend on your gut flora to survive, cattle depend on communities of bacteria in their rumens to digest cellulose and other complex compounds in grass. There is no doubt that the assault of high starch and grain diets, and antibiotics, of all things, weaken the ability of these animals to live well. Additionally, the prevalence of genetically modified grains, which are treated with herbicides like Glyphosate, assault the metabolic pathways of beneficial bacteria, making it even more difficult for the animal's internal ecosystem to thrive. We know that it is possible to create systems that mimic the natural tendencies of cows, which is to graze freely on diverse forages. I'm tired of seeing endless assaults on animal production altogether, simply because we see how dangerous the status quo has become, and we are either too tired or too lazy to find a middle ground. Just as the omega-3 fatty acid or the X-factor enzyme or the antioxidant will not alone make us healthy, there will not be one magic change to cure our system.

Of course, we cannot take our current beef consumption and pair it with a radically new production system and expect health, wealth, and triumph. And people are not going to quit eating meat, full stop. Indeed, many people do not have the luxury of cherry-picking what they eat, and cannot afford to eliminate this food and that food from their diet. We know that cattle eat grass, and we cannot. If we are to feed animals and people well, we should strive to find production systems, both of the backyard scale and of the Alan Savory scale, that favor this naturally efficient phenomenon.

Speaking of middle ground, I'd like to see research on the true nutritional differences between beef that is strictly grass-fed and beef that is pasture-raised with free choice supplementation of grain. In keeping with my philosophy of diet diversity, I'd argue that there is probably not much difference, in the fabric of a meat-eater's total nutrition, between eating responsibly raised grass-fed and responsible pasture-raised beef. I suspect the difference is largely one of taste, and there is not enough research at current to refute this claim.

If you choose to raise your cattle on grass only, look again to diversity. While you can simply turn an animal out into a fenced field and let it graze on what is growing there already, you may soon find that the quality of your pasture is declining. Just like you, animals eat what they like first and foremost, turn away from things that are less tasty and completely avoid the poisonous. Obviously, the plants the cattle avoid will be left to reproduce from seed, and will soon outnumber the best forage species in your pasture ecosystem. In addition, the pasture has its own intricacies (of course), like forages that only grow in wintertime, others that die off every year and some that harbor parasites. So the higher the diversity within your pasture (including the soil that supports the grass), the better you will be able to maintain it, and the better you will be able to feed your beef.

Begin to learn your grasses, such as bluegrass, orchard grass, fescue and timothy. Look for legumes (nitrogen-fixing plants) as well, which provide another kind of nutrition for your pasture's soil and your animals. Legumes include clovers, lupines, alfalfa and trefoil, among others. Once you've figured out what you have, you should seek to introduce more species, favoring a diversity of cool-season and warm-season forages, grasses as well as legumes, and plants that cattle will prefer (like orchard grass) versus plants they will eat but prefer less (like fescue). Note that these examples are taken from forage species that thrive in my area; if you are farming in a supremely different zone, you will need to choose from species adapted to your climate and growing season. You can drill in seed if you have the resources, or you can broadcast seed and see what takes. Depending on the size of your holding, you can choose to water and fertilize (organically, of course) your pasture or let the rains and the animals' manure provide your support. Whichever you choose, be sure to take yearly soil tests to determine your pasture's pH and overall nutrient availability. As we've

Rotational Grazing

Framers rotate animals between grazing areas for three reasons: to allow the forages in the pasture to regenerate; to ensure the best possible nutrition for the animals; and to provide removal periods in the life cycles of parasites, thereby increasing herd or flock health. The science behind this type of grazing management is based on our knowledge of the optimal stages in the growth cycles of grass plants, so we can stop grazing animals from eating those plants at the right times and let them regenerate. There are also optimal stages in the growth cycle of forage plants that offer the best palatability, bulk and nutrition for the grazing animal. To manage your herd using rotational grazing principles your pasture will have to be subdivided (using paddocks, as mentioned above) and you will have to move the animals on a regular basis. You must also make sure that the animals have access to fresh, clean water in each paddock.

For more reading on the foundations behind rotational grazing I recommend *All Flesh is Grass* by Gene Logsdon; further information on intensive rotational grazing, or mob grazing, can be found in *Holistic Management* by Allan Savory. I have noted additional books in the Resources section at the back of the book. ◆

seen, all things are connected. If the soil is not healthy, the grass will not thrive, and neither will the animals—or you.

If you choose to supplement with grain, the most typical options are barley or corn. Spent barley from breweries is a popular, recycled feed supplement for livestock in pastured systems, although producers should be careful about timing, as the animals find the barley very palatable when it is fresh but less tasty as it begins to ferment. I believe that barley in the earlier stages of fermentation can be extremely beneficial for the animal's rumen, and I would love to see research on this. The standard argument that all grain fed to ruminant animals is detrimental to their gut composition could use some educated, dynamic dialogue. Just as human's consumption of fermented foods can boost our immunity, hormone function, digestion and overall health, I'd submit that fermented grains can and may become a viable option for pasture-based feeding systems. Pre-sprouted grains are another creative option for feed supplementation.

Rationing is of utmost importance in grain supplementation, depending on the feed you're using. Too much spent barley, for example, can create vitamin and mineral deficiencies in the animal, while too much corn can adversely impact the pH of the rumen. You want to ensure that the animal's diet is at least 65 percent roughage or forage. In the highly industrialized systems, emphasis is placed on fast weight gain (as much as three to four pounds per day in the final 120 days of life), and steers can be fed up to 25 pounds of corn in a day. On your small homestead, these margins and timelines are not as important, and your production should aim to lessen animal stress. As such, supplementation should start slowly and remain reasonable. As little as three to five pounds of grain per day per steer is an OK starting place, especially if the animal has been on grass for most of its life and you are supplementing with grain for the last three to four months. Watch the intake, and increase it if you feel inclined. Again, if there is constant, quality pasture, the grain is merely a supplement, not a necessity. It is extra energy, which will turn into useful (and delicious) fat.

In compiling your feeding strategy, consider how your animals will overwinter. In a grass-fed system, the forages will slow down or stop their growth in wintertime, so you may consider supplementing with hay or stock piling forage by letting parts of your pastures rest and put on growth in the summertime, and then saving this for winter. You can also grow high-energy crops in holding areas for wintertime grazing. We used to do this year-round when we rotated cattle with pigs. Once the pigs rooted through a section of pasture, we would sow a fast-growing, high-energy grass, such as oats, barley, rye or sorghum (depending on the season). Once the crop put on sufficient growth, we would turn the cattle into it, giving them a shot of energy and nutrients before rotating them back to the diverse pasture.

This strategy leaves a lot of room for creativity for homesteaders with vegetable holdings. Perhaps you're choosing to cover crop a larger area of your garden for a season. You can then simply turn that cover crop into the soil, providing organic matter for the garden bed, or can choose to graze it, benefiting from the manure and urine of the animal in the garden area and donating the high-energy nutrients of that cover crop to your meat crop. The intricacies of your system and the end result you desire will determine your overall use of your land resources and your strategies for feeding your animals.

Lastly, your cattle will require a mineral and salt block. Purchase these at feed and seed stores, or refer to the Resources section for providers. Animals require macro- and micronutrients for growth and nutrition, and also for immune defense. Arguments against providing mineral and salt licks include claims that humans are providing synthetic supplements because they are forcing animals to live in a mineral-deficient ecosystem. These claims are false. Even in the wild, animals visit natural mineral and salt deposits, and are known to consume clay and dirt to provide themselves with nutrients.

Beef Butchery

The very first step in breaking an entire beef carcass is separating the animal into halves, and then quarters. To start with, I'd urge you to have your processor do this for you. He or she will first halve the carcass by sawing down the center of the backbone, a job few home butchers have the tools or space to do. The carcass is then quartered by straight cutting between the 12th and 13th ribs on each side.

The two quarters from the front of the animal are hereafter referred to as "forequarters." The back quarters are called "hindquarters." See the drawing at the beginning of this section for anatomical detail.

The Forequarter

Starting at the front seems appropriate, so we'll devote this section to the beef forequarter. From quarters, the next step is to break the carcass into "primals," the next largest portions. The beef forequarter has four primals: the chuck, the brisket, the rib and the plate. My recommendation is to have your processor section this forequarter into two pieces for you, breaking it between the 5th and 6th ribs. Each piece will have two primals: one portion will include the chuck and brisket, and the other portion will include the rib and the plate. If you're starting with the whole forequarter instead, your first step will be this same break, between the 5th and 6th ribs.

I had the pleasure of visiting with some of my meat friends in the great state of Virginia recently, and we will journey through beef butchery with them. Kilan Brown, head butcher at Pendulum Fine Meats in Norfolk, VA, will be our demonstrator model for the beef forequarter. As per the recommendation above, we will be working with two portions of the forequarter.

An important reminder as you read: butchery is both a science and an art. There is a right way to break the animal, but there is not just one way. Before you start, you must determine what you want out of the carcass, and cut accordingly. For example, you may choose to leave portions of the chuck intact for grind or roasting, or you may seam them out, as Kilan does, into steak portions. For instructional purposes, we've made decisions about what to do with the carcass, and we're showing you this approach only. As you learn beef musculature, you'll be able to stray from this approach and determine your own

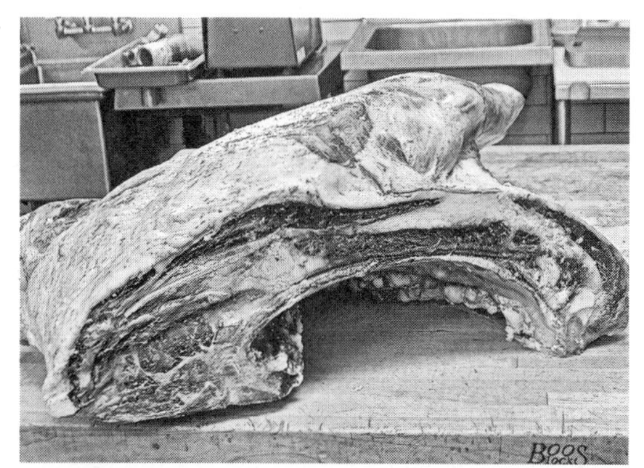
The forequarter from rib-end.

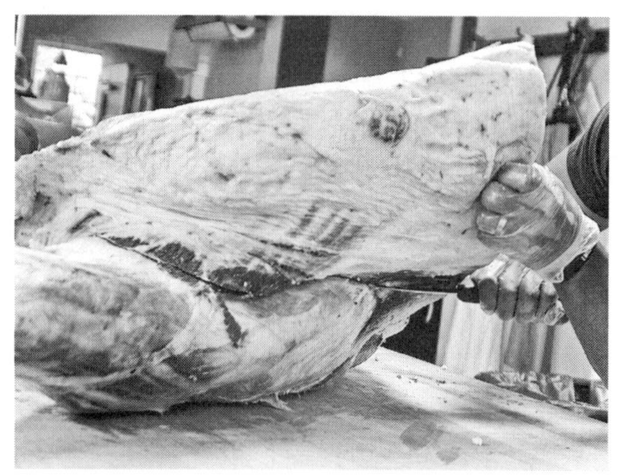
Kilan scores around the arm to help free the brisket.

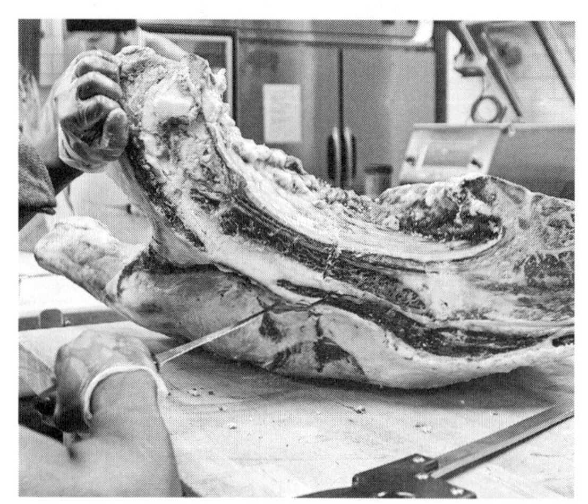
Bring the cut around the arm and meet with the mark on the ribs.

path. The awesome thing about whole-animal butchery is that you can be very creative with the carcass. You are the boss, and there are hundreds of possibilities!

The Chuck and Brisket Portion

Here's the portion, from the rib-end (end closest to the rib primal, which is not shown). Kilan's first move will be to remove the brisket. If your carcass has the neck attached, you can remove it first by finding the spot where the spine starts to curve back, at the 6th neck bone. Saw between the 6th and 7th cervical (neck) vertebrae, and finish cutting through the meat with your knife. You can debone the neck and roast it, or roast it on the bone.

To remove the brisket: Start by marking a line, beginning just underneath the 1st rib and parallel to the spine on the inside of the carcass, using the boning knife. This is where you will start to saw through the ribs.

Remember to saw through bone only. Stop sawing when you've reached the meat on the underside of the rib bones.

Next, move around to the other side of the carcass and use your boning knife to score around the arm. Follow this score mark around the brisket primal until you meet up with your saw mark on the end. You can see that Kilan's mark follows the contour of the arm until it meets his mark at the rib-end.

Once you've made these marks and sawed through the ribs, you'll use your knife to cut deeper, and slowly remove the brisket.

As Kilan works deeper with his knife, he begins to lift the brisket. You'll find that you use your hands a lot more than you think when butchering. Any time you can lift, manually pull, or use gravity to do the work, you will learn the natural seams and tendencies of the muscles to separate.

Once you've removed the brisket, it will still have some ribs and breastbone attached. You can remove these with your boning knife by inserting it just between the ribs and the meat and cutting close to the bone.

You can further trim up the brisket, but not too much. Halve your brisket to make one thicker piece and one thinner piece. Braise or smoke this muscle for best results, or make it into corned beef or pastrami.

Removing the foreshank: Kilan's next step is to remove the front leg, or foreshank, just above the elbow joint. He uses his knife to cut through the meat on top of the bone, then saws through the joint and finishes with his knife, to get through the meat on the other side. The shank is great for braising, as in beef osso bucco. For this dish, you'll crosscut the shank

Cutting out the brisket.

Deboning the brisket.

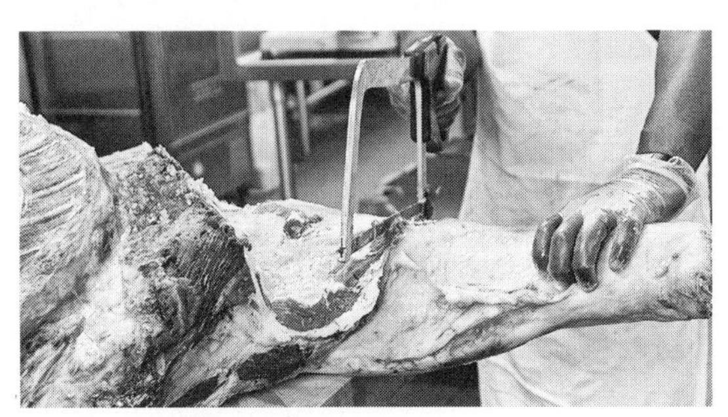

Kilan removes the shank.

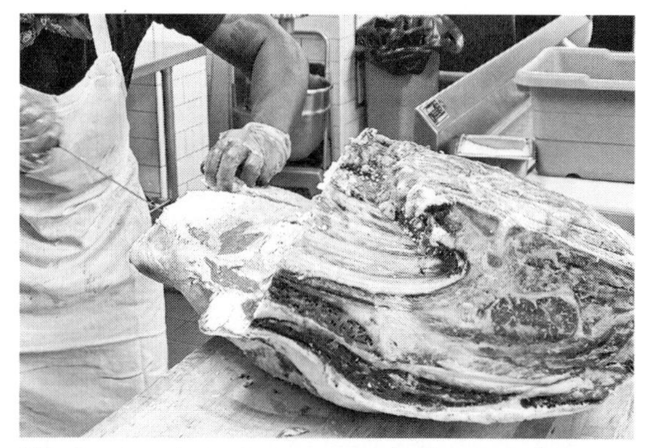
Kilan opens the arm to expose the humerus bone.

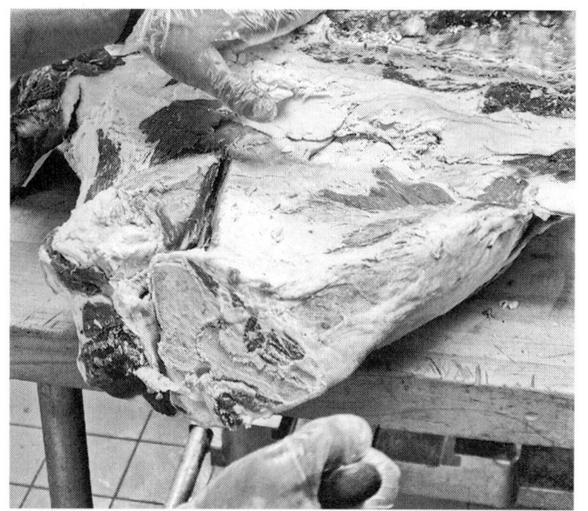
Keep the knife close to the bone and work around it to debone the arm.

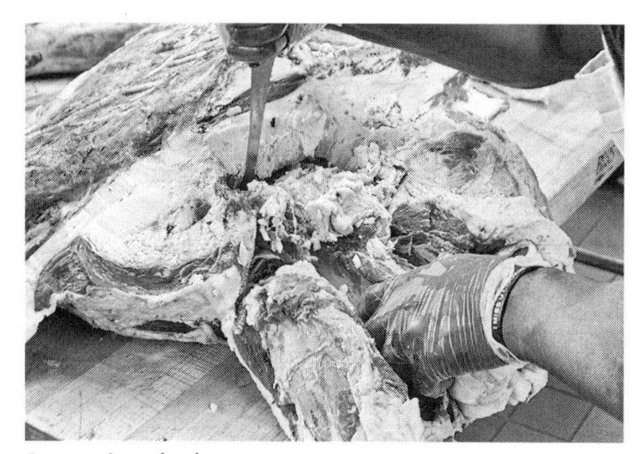
Removing the humerus.

into two- to three-inch rounds, with the bone running through the center. Alternatively, for some boning practice, you can debone the shank and tie up the meat for braising.

Removing the humerus: Kilan's next move is to remove the humerus bone, which joins the upper arm to the scapula or shoulder blade. To do this, he taps around with the tip of his knife on the underside of the arm until he feels the ball at the top of the humerus, where it fits into the scapula. Once he finds this, he makes a vertical cut toward himself, exposing the top of the humerus bone.

Once he has this cut, he begins to cut around the bone, keeping his knife as close to the bone as possible and letting the bone guide him. From the cut-end where the shank was removed, he can see the end of the humerus, which is another guide for opening the upper arm.

As you go, you may need to completely remove a section of meat and fat from atop the humerus bone so you can see better. This is OK. The fat and meat from the upper arm is perfect for grinding, so you don't need to keep it super intact. In the photo on the left, Kilan has removed a considerable bit of muscle and fat to expose the joint between the humerus and scapula, and get to the humerus more easily.

Be patient and continue cutting against the bone, peeling the meat back from it as you go. At the top of the arm, where humerus and scapula meet, you will have to knife under the shoulder meat to fully cut around the top of the humerus. Don't cut into the shoulder meat if you can help it; there are valuable muscles there in the chuck primal, which we will see more of shortly. Eventually, you will be able to lift the humerus out, once you've cut all the way around it. Admire the beautifully perfect-white joint area, untouched by blood. This is one of my favorite sights. If you see clear liquid coming from the carcass when you are splitting at joints, don't worry—this is joint fluid, which the body uses to cushion the bones. When you're finished, you should be able to see the socket of the scapula.

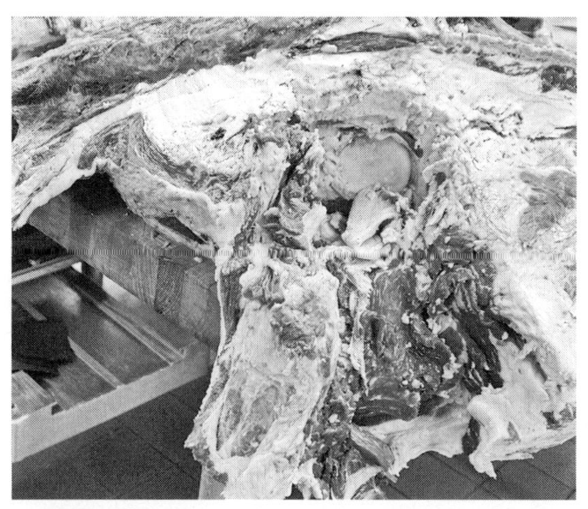

You should see the socket on the scapula once the humerus is gone.

Don't throw away the humerus! This bone can be roasted and used for making stock.

The chuck primal is best understood in two halves: the blade half (blade referring to shoulder blade) and the arm half. The subprimal associated with the arm half is the shoulder clod. The subprimal associated with the blade half is the chuck roll.

The shoulder clod: To isolate the shoulder clod, begin by flipping the chuck primal over, so you're looking at the outside of the carcass. This is a tricky step, as you are essentially peeling the shoulder clod off of the top of the scapula (shoulder blade). Due to the intricacies of the blade bone, this will take some practice.

Find the scapular spine and begin to trace it with your boning knife to isolate the shoulder clod.

Use your fingers, your knife and the visual of the end of the scapula (which you exposed upon removal of the humerus) to find the scapular spine, which is the pronounced ridge in the shoulder blade. You may also be able to

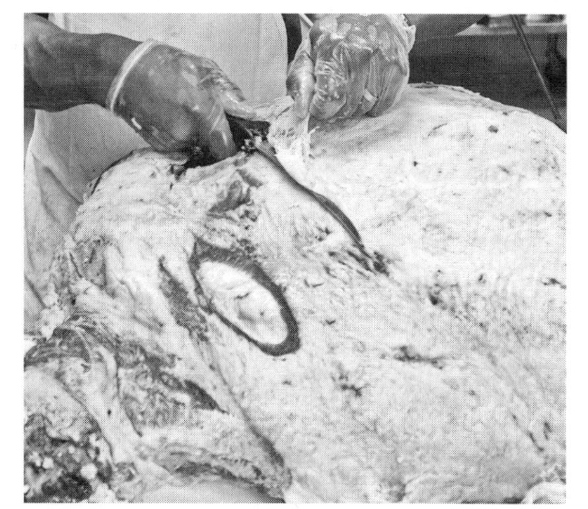
Removing fat to find muscle separations.

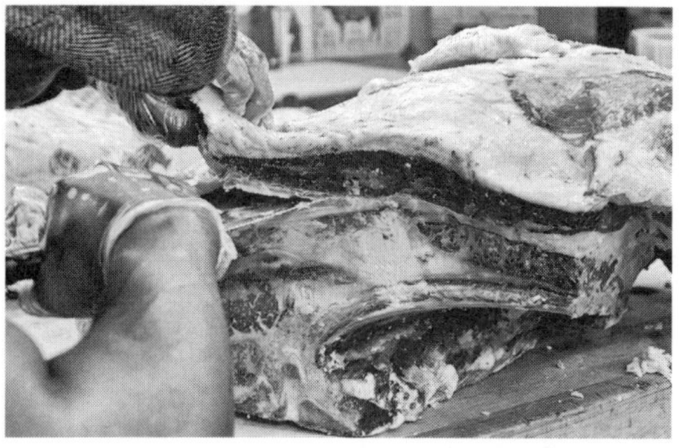
Continue around atop the ribs to isolate the shoulder clod.

Cutting the shoulder clod free; the scapula is underneath.

see a white seam that separates the muscles, but this takes a trained eye, and sometimes removal of some exterior fat.

Follow the seam around the top of the shoulder. At this point, you need to knife under the meat on top of the ribs.

Set the shoulder clod aside, in a cooler, until later.

Remove the mock tender: Nestled on the other side of the scapular spine from where you just removed the shoulder clod is the mock tender. Insert your knife along the scapular spine and follow the bone to remove it, lifting as you cut under it gently.

The mock tender is great butterflied, marinated and braised, or made into a stew. It's dubbed mock tender because it looks like a tenderloin but is not nearly as tender. Cook it accordingly, with low heat and some time.

Remove the scapula: Using the skills you gained boning out the humerus, follow the

Freeing the mock tender.

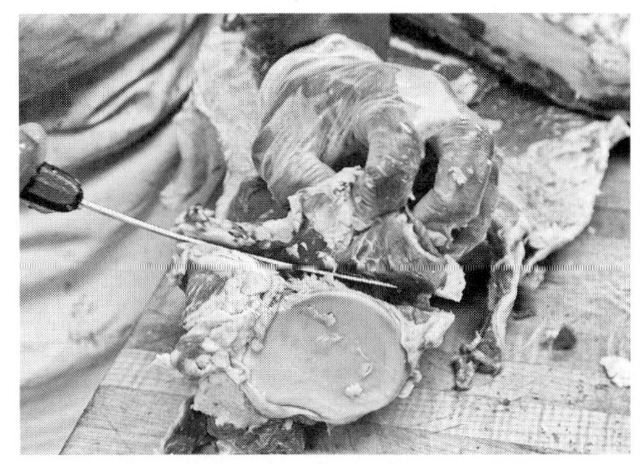
Cutting the splenius free from the scapula bone.

contours of the scapula with your boning knife to remove it. In our demo, Kilan removed it with a muscle still attached to its underside, the splenius, or the Vegas steak. This is more difficult to do, as it requires you to differentiate the splenius from other muscles in the chuck roll as you remove the scapula. You can also remove the scapula by itself, and then take the splenius off later.

If you're using Kilan's approach, once you've gotten the scapula off, get right up on the bone, starting at the socket joint, cutting as you lift, to pull the Vegas strip from the bone.

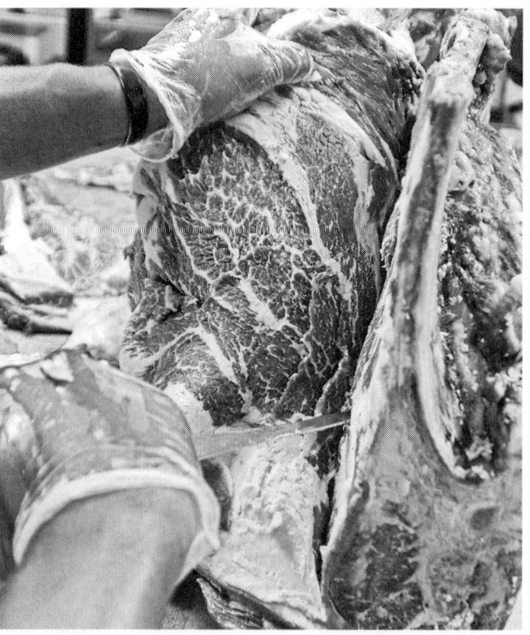
Freeing the Denver from the rib bones.

Removing the Denver: Now that you've taken off the scapula, you're looking right at the Denver, or underblade roast, an awesome cut of the shoulder that you can turn into beautiful steaks for the grill. Remove the underblade by following the ribs with your knife and pulling the muscle away from them.

Clean up the Denver and cut steaks. Measure your steaks using two fingers side by side

Cutting Denver steaks.

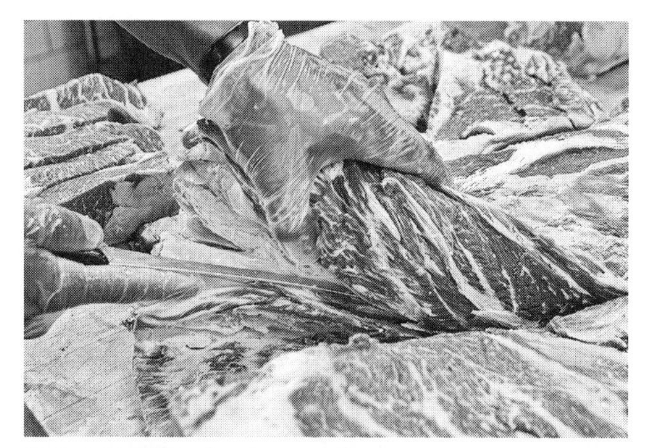

Removing the chuck eye roll.

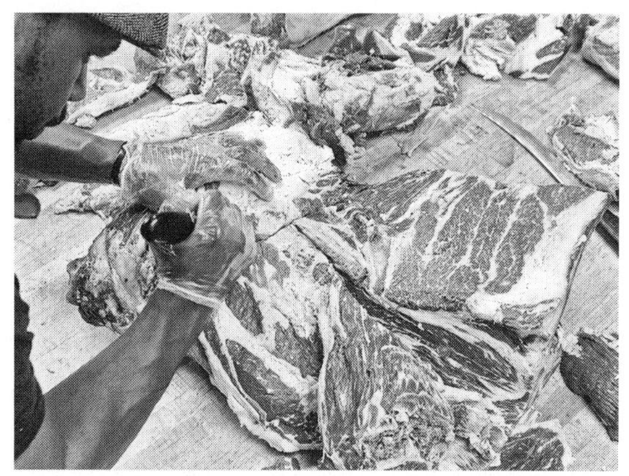

Marking the end of the chuck eye roll, the part closest to Kilan. The section of ribs, from which he removed the Denver previously, is also shown.

Removing the backstrap.

(great grilling thickness), and make clean, intentional cuts with your cimeter, first pushing all the way into the meat and then pulling back toward your body, once the knife's blade is even with the bench.

Pulling the chuck eye roll: The chuck eye roll is a prized group of muscles that run parallel with the spine. The seam for separation corresponds with where the vertebrae once attached to the ribs. Starting with your knife underneath it, begin to cut and roll it out, following the obvious seam.

Kilan did not remove the neck bones from the carcass yet, so he must mark the boundaries of the chuck eye muscles as they differentiate from the neck. At home, it may be easier for you to discern the muscle groups if you bone out the neck first, and then proceed with breaking down the chuck blade portion.

Now you can continue to undercut the chuck eye roll. Once you've pulled the chuck eye roll, you'll need to remove the backstrap, a thick ligament. Simply undercut it, lifting it as you go.

Once you've removed the backstrap, you can cut up to three chuck eye steaks. You can also make fewer cuts to produce boneless chuck roast portions. More recently, butchers are isolating each of the four muscles in the chuck eye roll, as these are some of the tenderest muscles in the carcass. Follow the seams to achieve this, and cut into steaks.

At this point, you'll have a fair amount of trim, which is the proper term for extra, miscellaneous portions of both muscle and fat that can be used for grinding. You'll also have a lot of bone, which you can clean up later, when you have more time to pull any remaining meat from it for grinding. If you haven't been organizing as you go, try to create some order by putting all the bones, fat and muscle in three separate piles. If you have offers for help, have friends begin cubing your trim into roughly two-inch cubes for grinding. Keeping the fat and muscle separate will allow you to easily weigh it later. This will be essential for getting the correct ratio of lean to fat for good ground beef or beef sausage.

Learn to differentiate between good fat and bad fat for grinding. Good fat is white and firm; bad fat, which will ruin your grind, is yellowish and flabby. You'll also want to exclude any glands you come across, which are yellow or off-white and round, and any dry age on the outside of the carcass. The bad fat, grind and mold will turn your ground beef rancid, so carefully pick through your remnants to divide and conquer.

The first cut into the shoulder clod.

Back to the shoulder clod and finding the flat iron: Within the shoulder clod subprimal you'll find three cuts: the flat iron or top blade, the petite tender and the ranch or underarm roast.

Remove the shoulder clod from cold storage and place it on the bench so you're looking down the length of it. You'll notice a thinner portion and a thicker portion. There's a seam separating these two portions, almost in the middle of the shoulder clod. That is your starting place. Begin your cut down this seam to

Freeing the flat iron.

Find the seam at the end of the flat iron.

begin isolating the top blade or flat iron. It's the flatter of the two portions.

For true top blade steaks, you'd simply cut steaks from the entire top blade portion. More often, the muscle is separated at the interior seam to produce two halves. If you've heard of a flat iron steak, this is the technique used. Pull the fat from the underside of the top blade. You should be able to at least start this process by hand, and then use your knife to assist you as you go.

Now you'll need to find the seam at the end of the flat iron, starting at the thicker side. It's about halfway through. You'll begin here and cut laterally through the muscle to expose the steak underneath. This step will be challenging. Take your time.

Once you've finished that cut, flip the steak over and begin cleaning off the other side.

Once removed, you'll have beautiful, long, flat iron steaks. You can divide the muscles

Carefully remove the top muscle to free the flat iron below.

into portions, or simply halve them and cook larger portions.

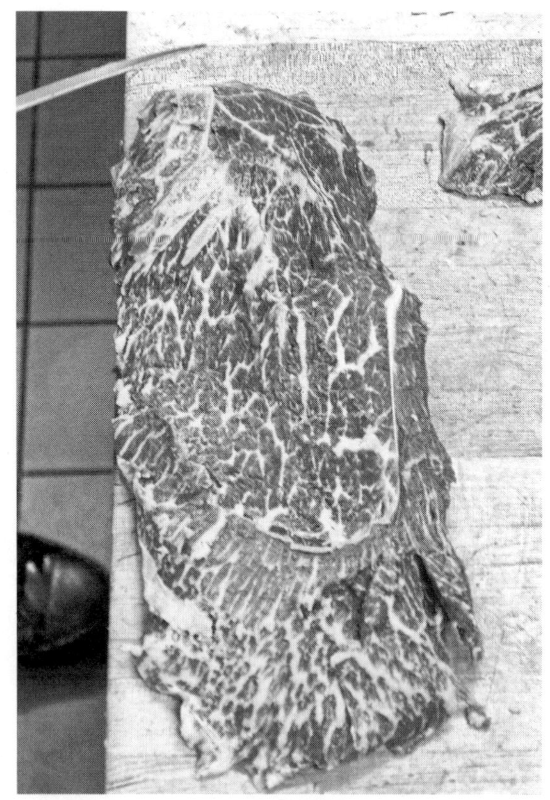
The flat iron steak.

The petite tender: Go back to the other portion of shoulder clod. You may need to remove some fat here to see the petite tender, a muscle that runs alongside the edge. Pay attention to muscle grain. Where you see the direction changing is where you want to look for a seam.

Remove the petite tender by lifting and gently cutting with the seam. This muscle is versatile and can be roasted or broiled whole, or cut into medallions and grilled. Cook it medium or medium rare.

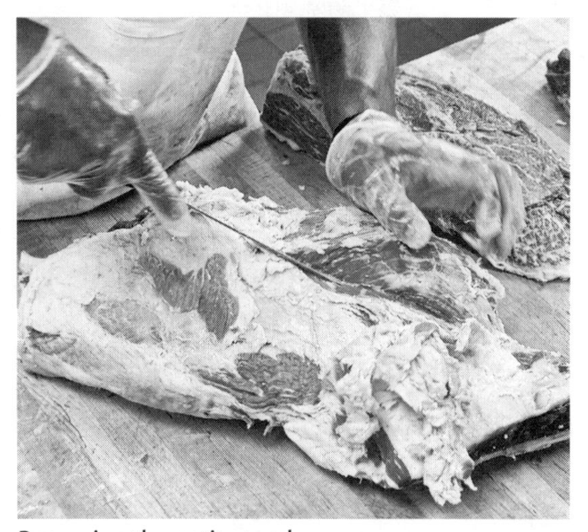
Removing the petite tender.

The ranch or underarm roast: At this point, you're left to isolate and clean up the ranch, a popular specialty cut that makes great grilling steaks. Start by cleaning up the area where you removed the petite tender, beveling fat and trim. On the animal Kilan is handling, there was considerable fat, so he had to remove quite a bit.

Now flip the piece over and begin to remove the fat cap. As you work at this, you will begin to see the muscle grain in the ranch. Let this guide you as you remove fat from the top and either side of the muscle.

This will take a good bit of cleaning, and you'll end up with a good amount of trim for ground beef or beef sausage. Let the muscle grain guide you. The general diamond shape of the muscle is a loose goal.

The ranch's rough diamond shape.

Pull the cap muscle off, starting at the seam and lifting from there.

Clean the inedible silverskin from the ranch, and any extra fat, by slipping your knife underneath it and carefully shaving it off.

Clean up by beveling the edges and squaring off the sides. Now you can cut steaks, using the two-finger measuring method we learned earlier. Make sure your knife work is clean and deliberate so you don't end up with ragged steaks.

Opposite is a shot of everything we pulled from the chuck and brisket primals, which is everything from the neck of the animal down through the 5th rib. Once you've sealed and stored all this, you're ready to move onto the second portion of the forequarter: the rib and plate section.

Removing cap muscle to reveal ranch.

Removing silverskin.

The chuck and brisket cuts. From left, top row: halved brisket, trim for grind, meat from upper arm, the foreshank, and the other half of the brisket. Second row: Flat iron steaks, chuck eye and boneless chuck steaks, the mock tender. Third Row: Ranch steaks, the chuck flat (part of Denver), petite tender, Vegas strip, boneless chuck steaks, denver steaks.

The Rib and Plate Portion

The rib primal and the plate primal make up the section between the 6th and 12th ribs of the animal. Kilan's first step is to remove the outside skirt steak, which is an obvious, easy-to-remove muscle that hangs (like a skirt hem) and runs the length of the ribcage. To remove it, simply find the seam of fat between the edge of the skirt and the ribcage, and separate from there, lifting the skirt as you go.

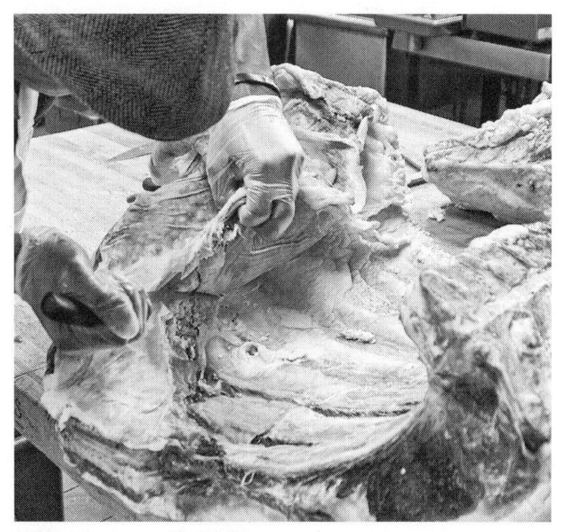
Kilan removes the skirt.

Set the skirt steak aside, or trim it up now. It will take a bit of cleaning. Just as in any trimming, undercut the fat with your knife and

Kilan marks his line for separating the prime rib from the plate.

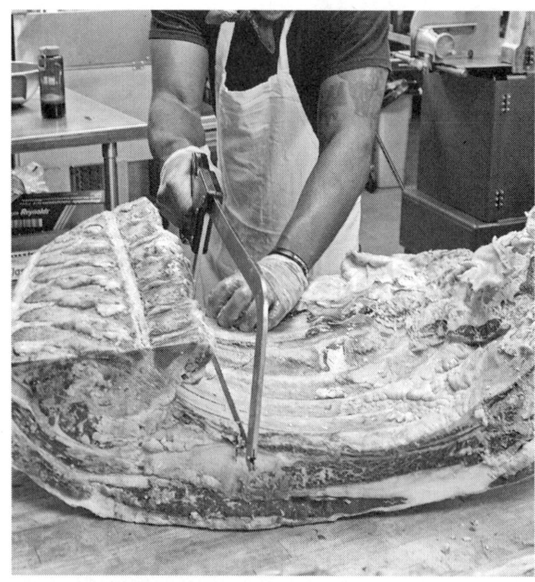

Kilan saws through the rib bones. He will finish this cut with his knife.

ease it off. You can see a photo of the cleaned skirt steak at the end of this section.

Next, it's time to separate the rib primal from the plate. Score a mark about one and a half times as long as the eye of the loin (the meatiest part of the ribeye) to determine your cut. Use the blade of your knife to help you measure (see p. 137 in the Pork Butchery sections for visuals).

Kilan marks with his knife and cuts through the end meat before he begins sawing. Saw through the rib bones only, then use your knife to cut through the meat below to finish the separation.

We'll work with the rib primal first. The first step is to remove the plate meat from the outside of the bone-in prime rib. You can begin this process using your hands, and simply separate beginning at the corner. You'll need assistance from your boning knife as you go. Cut the thinner plate meat into strips for braising, or contribute to the grind.

You have a couple of options for your next step. You could cut bone-in ribeye chops, by standing the prime rib on end and sawing off the chine or spine, removing feather bones, and then cutting steaks. Kilan is taking another option, working some extra magic on the prime rib to get more meat for his case. First, he'll bone out the prime rib. You're probably getting familiar with this process by now. Cut just atop the bones and begin to separate bones from meat as you go.

Once you've separated the bones from the

rib meat, you can begin to isolate one of the best cuts on the beef carcass: the ribeye cap. This is an exceptionally tender muscle that you can take off without sacrificing your ribeye steaks.

Turn the boneless ribeye so that the inside is on the bench and the loin is closest to you. The ribeye cap runs along the top of the loin. You can see it more easily if you look at the ribeye longways. You'll see the eye of the loin, which is the center of your ribeye, and then a seam of fat, and then a thinner muscle before you get to the outer fat cap. That thinner muscle is your ribeye cap. To isolate it, begin trimming fat from atop the loin until you can discern a seam. Once you've found it, you can begin to roll the cap off, that is, pull back on it as you cut under it with your knife.

Once you've removed the cap, you'll need to clean it up considerably, removing fat and

Kilan removes the plate meat from the outside of the prime rib.

Deboning the prime rib.

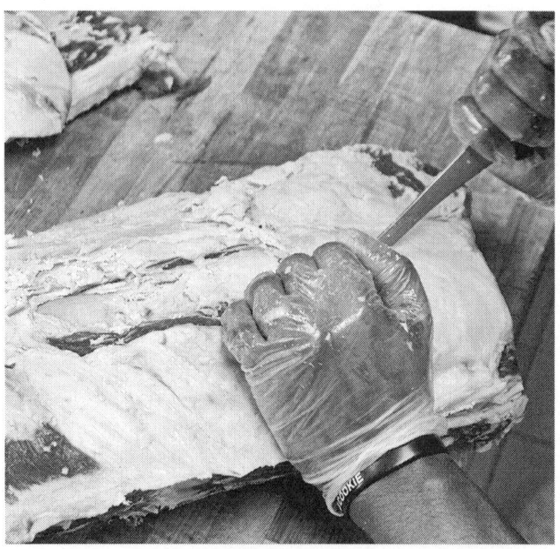

Kilan begins to take off the ribeye cap steak.

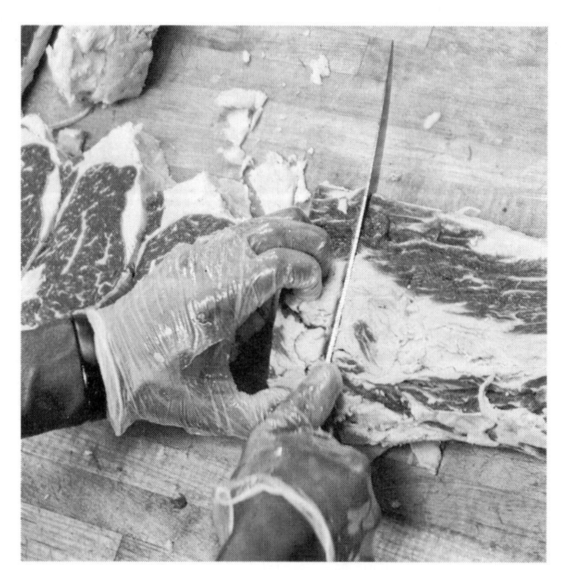
Kilan cuts boneless ribeye steaks.

backstrap. When you're done, it should look long and flat, with its beautiful muscle grain running the length of the steak.

Kilan will cut the cap into portions for his meat case, but at home, you may choose to leave it whole.

Next, you're ready to cut boneless steaks from the ribeye. Keep the ribeye with the loin up and the inside on the bench. Depending on the animal and the processor's aging process, you may need to remove a thin slice from the end of the ribeye that may have mold or oxidation on it. Then measure, using the two-finger method, and make clean cuts to produce your ribeye steaks.

Time to move on to the plate primal. This is your opportunity to create short ribs. If you choose to do this, saw off the thicker plate meat from the edge of the primal. You can use this meat for grind or curing. Then, with the short plate section remaining, crosscut again with your saw, then portion into short ribs by cutting between the ribs with your knife.

Instead of making short ribs, Kilan is going to bone out the plate section to produce an entire boneless beef plate, which is what you'll want for making beef bacon.

The process Kilan uses for deboning the plate is a bit different, and can be applied to other primals where you face the task of removing ribs, such as in the chuck roll. This method also lends better yields, and cleaner bones. Cut in between the ribs, then undercut them from the end, lifting the bones up as you go. This will leave the most meat on the plate.

Kilan will then halve the boneless plate, to make more manageable portions for curing and smoking.

In the photo opposite, you'll see all the cuts we've produced from the rib and plate primals.

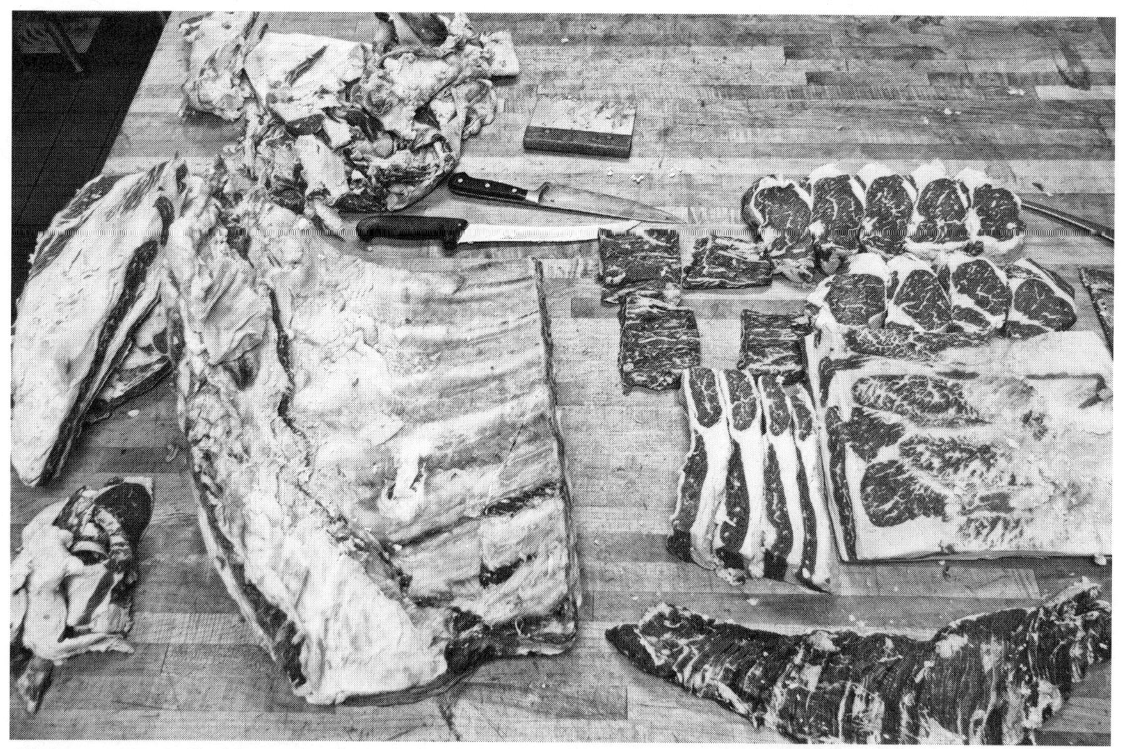

All the cuts we pulled from the rib and plate primals. From left, at top: trim for grind. Second row: deboned plate meat, rib bones, ribeye cap steaks, boneless ribeye steaks. Third row: boneless short ribs, boneless plate meat. Fourth row: skirt steak.

The Hindquarter

The beef hindquarter is everything past the 12th rib, and includes four primals: the loin, the flank, the sirloin and the round. I had the pleasure of working with the fine folks, including Matt Halpern and Tanya Cauthen, at Belmont Butchery in Richmond, Virginia, to break down a hindquarter. We'll also see some more of Kilan in this section, as he seams more muscles out for his case, while Tanya at Belmont tends to keep primals whole in her case, and then cut to order. We'll also see a little bit of Matt Helms, at the Chop Shop Butchery in Asheville, NC.

If you feel confident in the skills of your processor, I'd urge you to have your hindquarter broken down for you so that you receive the flank, loin and sirloin in one piece, and the round with the hind leg as a second piece. The only danger here is that your processor could split

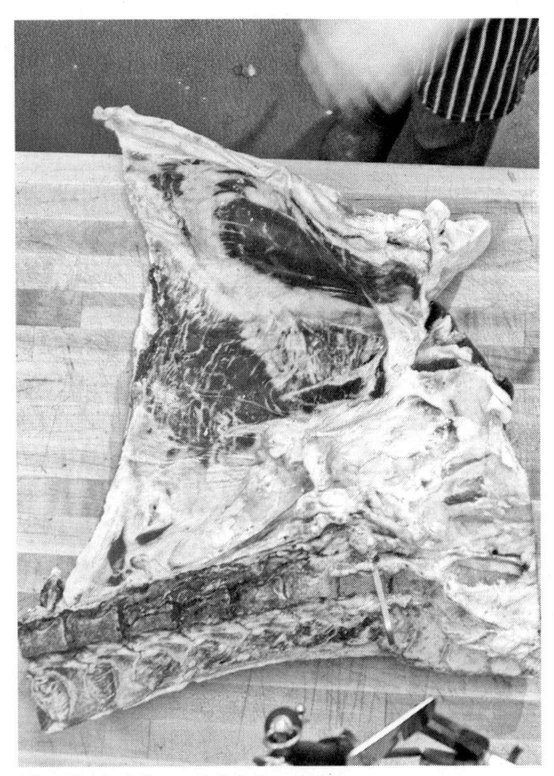

The flank, loin and sirloin section.

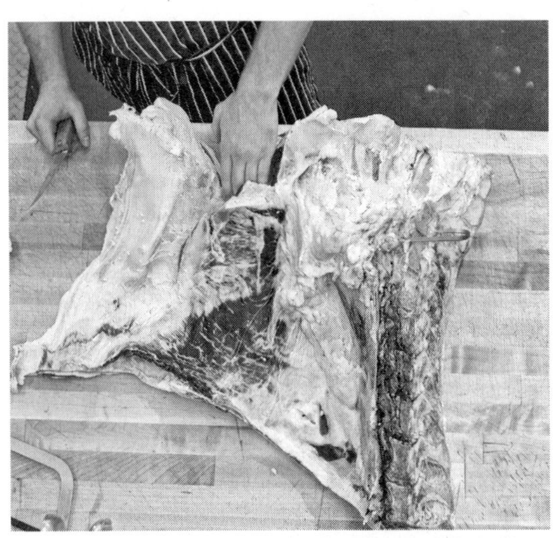

Matt prepares to remove the first flap. His hand is between the two flap pieces, to demonstrate their placement on the carcass.

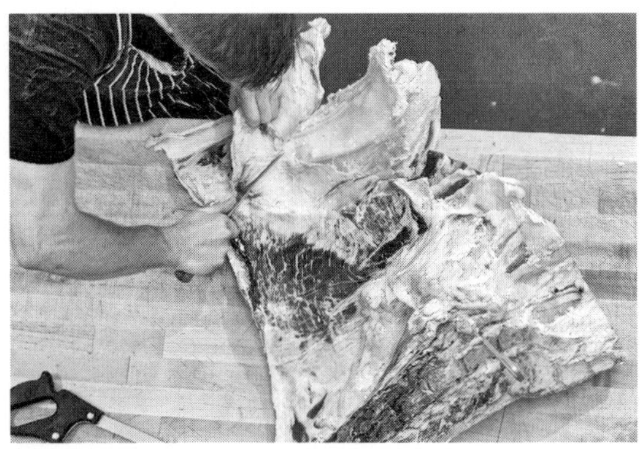

Matt removing the flank.

the round off in a way that damages the tri-tip muscle, which sits on the outside of the leg, so make sure to mention that you'd like the tri-tip removed first.

The Flank, Loin and Sirloin Section

Here are the flank, loin and sirloin primals at Belmont Butchery.

Removing the flank and flap meat: Matt will start by removing the flank steak, the large, flat, dark muscle opposite the spine. He does this by cutting along the obvious seam and pulling back with his free hand.

The flank will require a considerable amount of trimming before it looks like the flank steak you're used to seeing. It's enclosed in a lot of tissue, fat and sinew, which will give you a great opportunity to practice your knife skills.

Matt's next move will be to cut out the first piece of flap meat. There are two, stacked on top of one another. In this photo, Matt's hand is in the seam between the first and second flap, sometimes referred to as the inside and outside flap meat. This is the source of one of my favorite cuts, the *bavette* steak. Flap meat is great marinated and then cooked over high, dry heat.

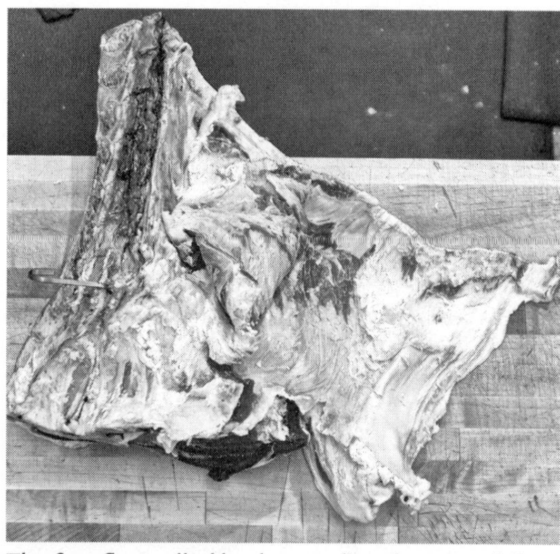

The first flap pulled back, revealing the second flap underneath.

You can see the edges of the flap meat because it is surrounded by fat and membrane. You should not run into much confusion when removing these pieces. When removing the second flap, you'll need to cut along the edge of the 13th rib to free the muscle.

At this point, we'll move back over to Kilan's block to further work the loin section. Tanya and Matt will leave the tenderloin, short loin, sirloin and second flap intact as one piece for their case. This allows them to fabricate cuts depending on what the customer wants. For example, if someone wants porterhouse or T-bone steaks, they can do that, or if someone wants whole tenderloin, they can pull that muscle out later.

Removing the tenderloin: The tenderloin is nestled up under spine, between the 13th rib and the pelvis. It's the tenderest muscle on the carcass, because the animal doesn't use it much. The other side of that coin is that muscles rarely used have less flavor. You have two options: leave it in and fabricate bone-in steaks that include some tenderloin, such as porterhouse and T-bone; or remove the entire tenderloin to produce filet mignon or use it whole. Keeping your knife close to the spine and serosa that surrounds the tenderloin, carefully remove the muscle, beginning with the tail of the tenderloin, which is the thinner end closer to the 13th rib, and ending with the head of the tenderloin, which is actually closer to the posterior (tail-end) of the animal. This intricacy can be confusing without a bit of explanation. As you approach the head of the tenderloin, you will probably have to cut and

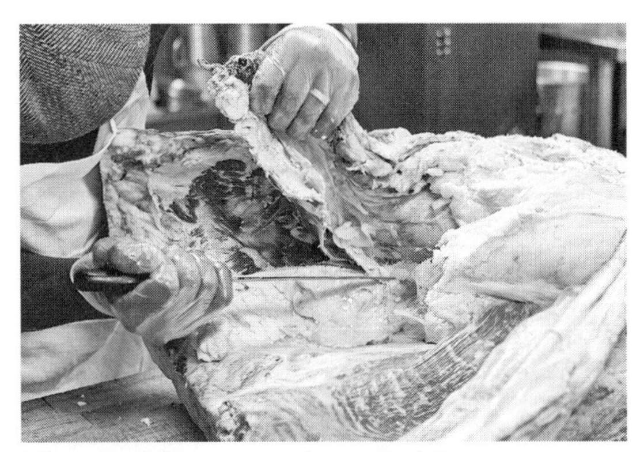
Kilan carefully cuts out the tenderloin.

scrape away the suet. This is a special type of fat that surrounds the animal's organs; you'll be able to differentiate it easily from other types of fat by its dry, crumbly texture. This is the fat you'll want to save for making tallow.

If I'm going to remove the tenderloin, I like to leave the fat on. There is something hugely satisfying about a fat-on, whole tenderloin, tied up and lying on a tray. Most people don't realize this is even an option, because we're so used to seeing little round tenderloin medallions with no fat at all. But if you're going to isolate the least flavorful muscle on the animal, some fat may help. You can always remove it after cooking, if you so choose.

Removing the tri-tip: The tri-tip is the cut from a muscle responsible for moving the thigh away from the midline of the body. It is thin, flat and superficial (on the surface of the carcass), spanning the hip and the leg, so it is best to remove it before separating the short loin from the sirloin. For this task, we visit with Matt Helms of the Chop Shop in my hometown of Asheville, NC.

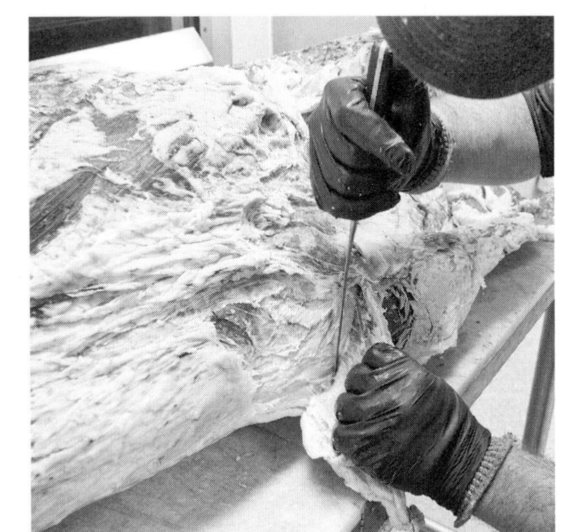
Matt begins to remove the tri-tip, beginning on the sirloin.

The tri-tip is so named because it has three corners: two on the hip and one on the leg. Cutting it takes practice, as some of the seams can be hard to find. To start, Matt faces the inside of the hindquarter and, starting at the sirloin, begins to peel back surface fat until he finds the corner of the tri-tip muscle and can begin to remove it.

He'll continue to pull and separate the muscle, moving toward the bench.

Now, flip the carcass over. You'll find the next corner of the tri-tip at the natural point where hip and thigh join. In this photo, you can see the valley between hip and thigh; and Matt is holding the tri-tip, looking under it to ensure clean cuts.

With this much of the muscle isolated, you can follow seams and watch muscle grain to discern the third corner of the tri-tip on the leg. It will not be a perfect triangle.

Matt continues to remove the tri-tip.

Separating the sirloin: Next, saw through the spine to separate the sirloin from the short loin. The sirloin primal is basically all the meat between the spinal column and the hip. To isolate the sirloin, Kilan at Pendulum Fine

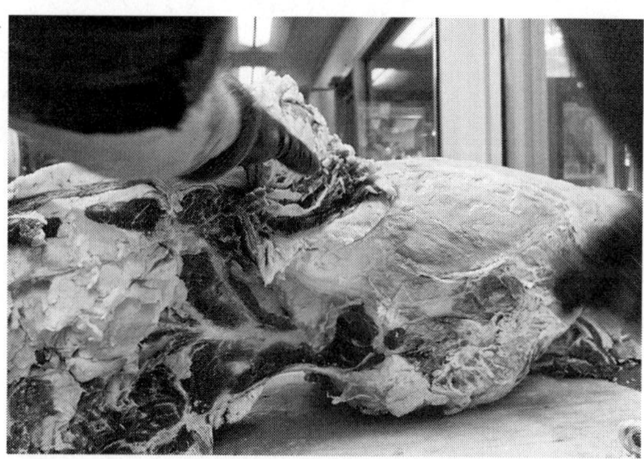
With the hindquarter turned over, Matt continues to isolate the tri-tip.

Loin Vocabulary

If all this talk of loins is confusing you, you're not the first. *Loin* meat is any meat that runs along the spinal column. The *short loin* runs between the last rib and the pelvis. *Sirloin* is meat between the pelvis and the hip (the word comes from *surloine*, which is French for "on the loin"). The *tenderloin* is a specific muscle within the loin. Most people are so familiar with tenderloin that they go into butchery thinking this is the only time anyone will be using the word "loin." Nope. An understanding of loin meat, how the sirloin differs, and knowing that the tenderloin is not the only loin will really help you understand anatomy better. ◆

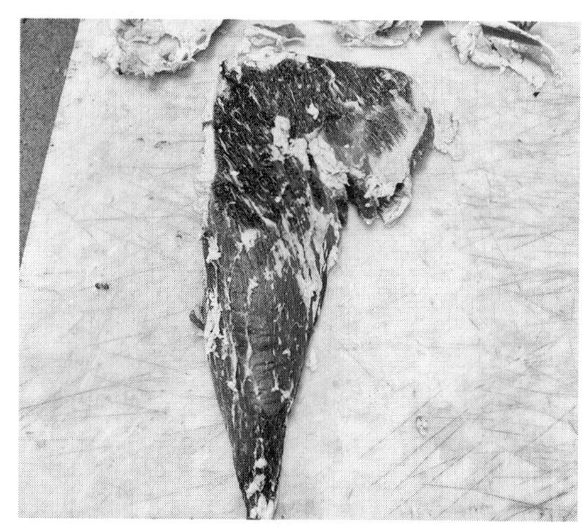
The cleaned tri-tip. Note the irregular shape.

Kilan saws the sirloin from the short loin. You can see the 13th rib sticking up on the right. Take that off, now, too.

Kilan cuts boneless NY strip steaks after deboning the short loin.

Meats places his saw at the second-to-last vertebrae before the spine begins to curve at the pelvis. He will also saw off the 13th rib, which remained after he removed the flap meat.

The New York strip: Kilan's next move will be to bone out the short loin to produce a boneless New York strip. From this, he will cut steaks. From your practice on the rib section, you should be plenty familiar with undercutting bone. Once you get the bones out, you're ready to cut boneless NY strip steaks.

Next, we'll deal with the sirloin. Within the sirloin primal, Kilan will isolate the sirloin cap and several sirloin center steaks. The first step is removing the pelvic bone.

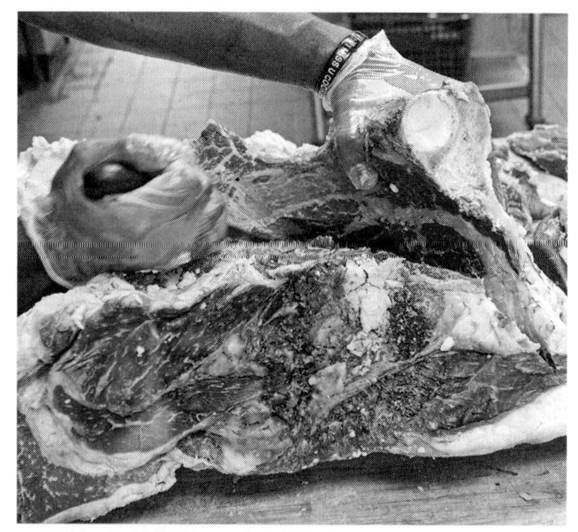

Kilan removes the pelvic bone.

Removing the coulotte: The coulotte is the cap to the sirloin. It is easily separated via an obvious seam. It is a surface muscle, so you'll also identify it by the nice layer of outside fat and dry age it tends to boast. Begin removing the coulotte by hand, then use the knife to help cut it away, pulling back with your free hand.

Once the coulotte is removed, you can trim it up, mostly to square it off. Don't remove too much of the fat cap, as this is one of the coulotte's signatures. When you're ready, cut steaks.

Removing the top sirloin center: Once you've removed the sirloin cap, or coulotte, start by trimming up what you have remaining from the sirloin primal. There will likely be a bit of age on the end of the sirloin, which you can shave off, and some fat that you can get rid of to round out the muscles. Now you should be

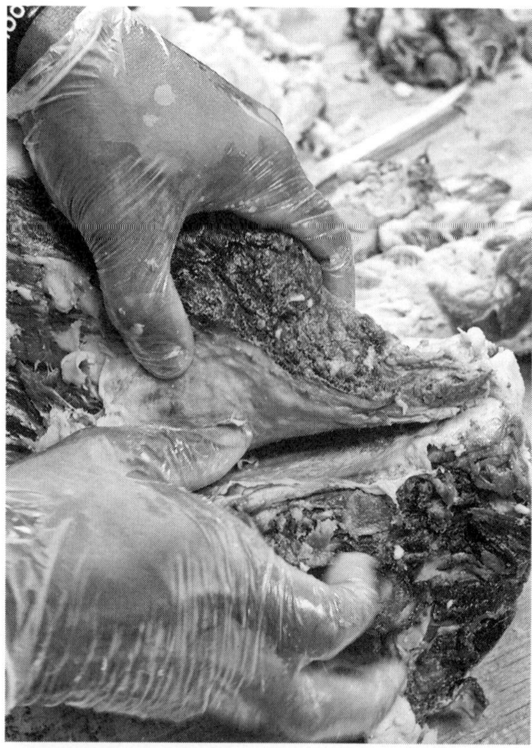

Kilan begins removing the coulotte by hand, via a prominent seam.

Removing the baseball.

Kilan cutting sirloin steaks.

able to see two muscles: a small, round muscle and a larger muscle. The smaller round muscle is often referred to as the baseball muscle. You'll be able to discern the seam, and follow it while rolling it out with your free hand. You can leave the muscle whole for roasting or cut it into medallions for nice, single-size steak portions, often called top sirloin filet steaks.

You're left with the larger of the top sirloin center muscles, which makes nice, sizeable boneless steaks. The only work left is to trim up and cut sirloin center steaks.

All the cuts from the flank, loin and sirloin. Top row: flap meat, pelvic bone. Second row: flap meat. Third row: flank steak, tenderloin, tri-tip (which has been compromised in this photo; note that one whole corner is missing), coulotte steaks, baseball/top sirloin filet steaks, lean trim. Bottom row: boneless NY Strip, NY strip steaks, boneless sirloin center steaks, suet.

The Round

The muscles in the beef round are sometimes the most perplexing from a culinary perspective. The animal uses these muscles a lot, so they are very lean and the least tender of all, without much fat to help out. We'll isolate the main muscles with Matt at Belmont Butchery, who was planning to turn most of the round meat into hot dogs.

Matt starts by removing the hindshank. He identifies his mark just above the knee joint, cutting the meat with his knife and using the saw to get through the bone.

The shank is great for braising, bone in or out. The Achilles tendon is also quite useful. A friend of mine made a bow drill out of the Achilles and anklebone of a cow, and many archers back their bows with strips of Achilles tendon.

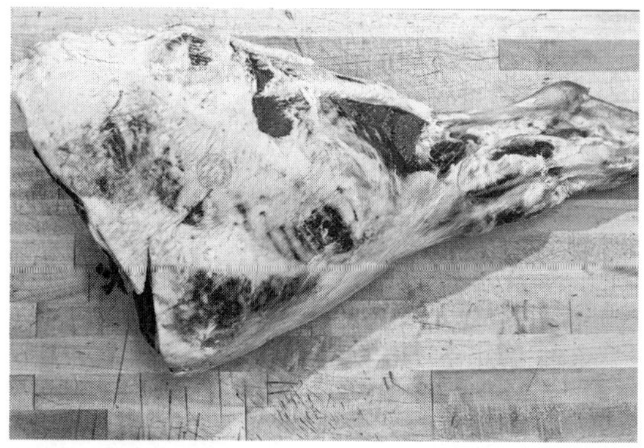

Here's the round.

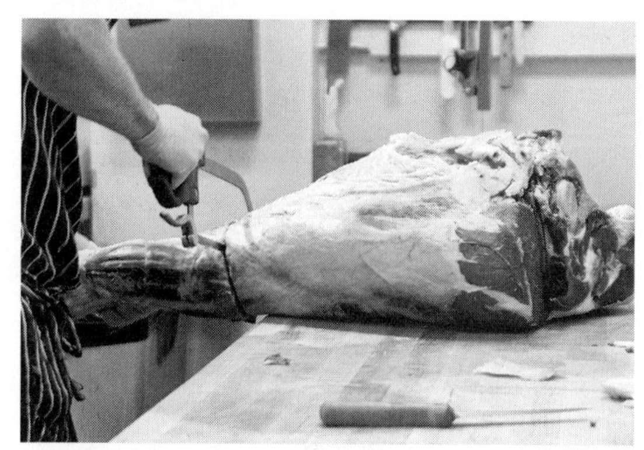

Remove the hindshank.

The top round: The top round is large and somewhat flat. This muscle is also called the inside round, because it is on the inside of the animal's leg. You may need to trim some of the fat to discern its seams. It can be tricky to isolate, so take your time and let the seams and muscle grain guide you. If you don't plan to grind the meat from the top round, cut it into steaks and try London broil, or use it in a stew.

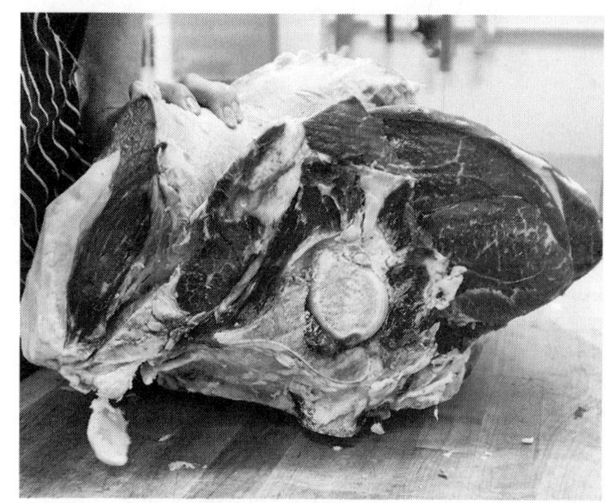

Matt's hand is pulling back on the top round at the seam. Under it, you can see the circular end of the eye of round, which you'll remove next.

The eye of round: this is a circular muscle in the center of the round. Many butchers loathe it, but I really like it. It's not the most flavorful or tender or fatty, but I like it for roast beef

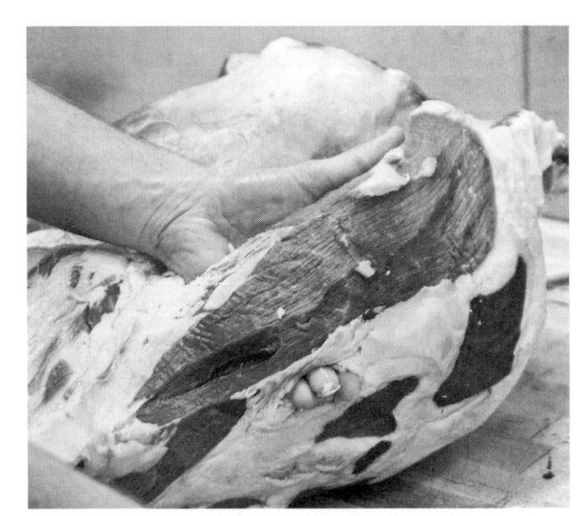
Matt shows off the eye of round.

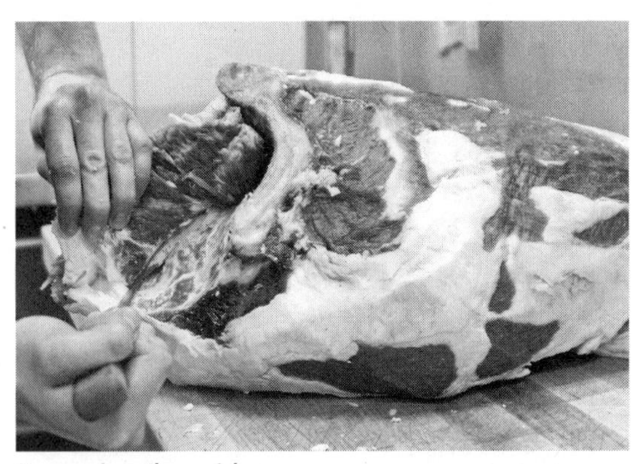
Removing the spider.

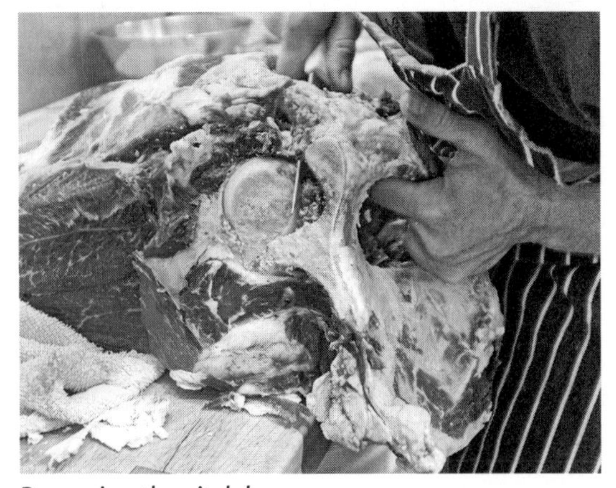
Removing the aitch bone.

and beef jerky. Kilan makes carpaccio out of it, which I think is brilliant, as this is typically done with the tenderloin.

Picking out the eye of round is easy. From the end of the round, it's the circular muscle, tapered on either end. From the side, you can see the decent seam that differentiates it from the bottom round. In the photo below, Matt has his hand around the eye of round, with his fingers poking through the seam underneath it.

To remove the eye of round, follow the seams and roll it out with your free hand, while cutting underneath it.

The spider: The spider is a lovely little muscle that sits inside the hip bone, or aitch bone. It's the oyster, if you will, and most butchers eat it for lunch, so you won't see it often. To remove it, simply cut against the bone and pull the spider from its concave home.

The bottom round: Next, you'll remove the bottom round. Start by taking out the aitch bone, carefully cutting under its contours to pull it away from the meat.

The next step is to lift the bottom round to cut at the seam under it. Follow through until you've removed it completely.

The knuckle: Next, follow the seams to remove the knuckle, which is largest quad muscle (the group between the hip and the knee). The knuckle can be seamed out into smaller steaks, sometimes referred to as sirloin steaks, but this bugs me, since it's in the round. At my shop, our favorite thing for the knuckle was a good spicy rub and a trip to the smoker. We'd sell smoked beef as deli meat, or, as my brother likes to eat it, sliced steak-thick for breakfast. He's a country boy.

To remove the knuckle, simply follow the seams. There are some seriously delineated muscles at this point, since you're right near the bones and joints of the leg. In the photo middle right, Matt holds the knuckle away from the leg so you can see its place easily. Many people choose to begin removing the femur at this point, and then pull the knuckle off.

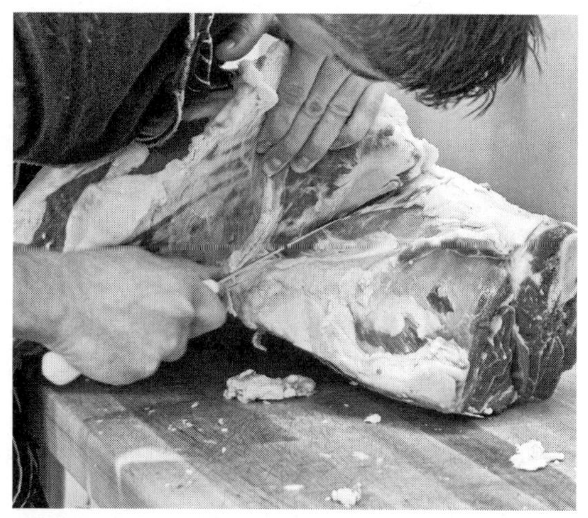

Matt removes the bottom round

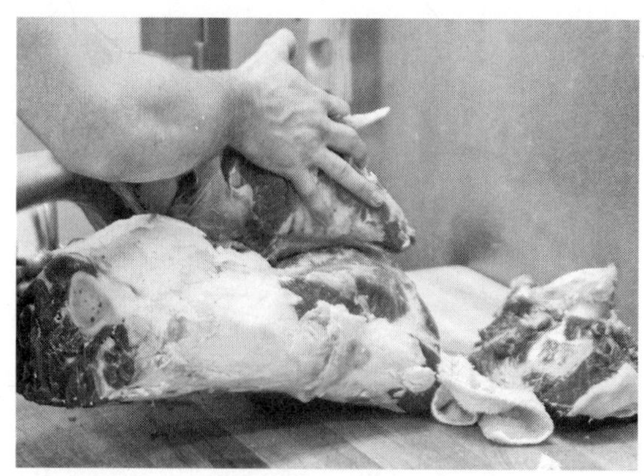

Remove the knuckle by following the obvious seams in the leg. Once removed, it looks like a dome: flat on the bottom, rounded on top.

The rat: Yes. We call it the rat, and it is also known as the heel. This is the last muscle left, just above the Achilles tendon. It's good for grinding, really, although I've heard you can seam out the muscle in the very center for decent eating. I haven't tried it yet.

You could take it further, seaming out muscles from the top round, bottom round and knuckle. Follow the seams and experiment!

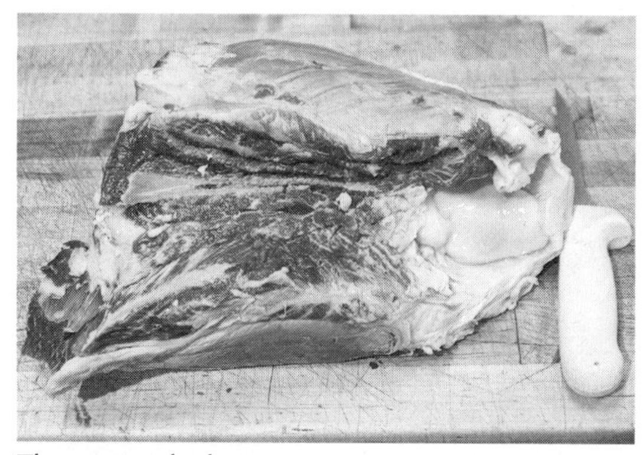

The rat stands alone.

Cooking With Beef

Beef is special. It's one of my favorite meats to cook with. All meat has its unique and special qualities, for sure, but beef holds a place in my heart. It seems like a prize every time. I appreciate the velvety quality of beef fat, and the clean, rich flavor of beef muscle. I love strong flavors, so I am thankful for beef in that it can stand up to some of the most colorful-tasting herbs, veggies and sauces.

All the muscles pulled from the round. From left, clockwise: the femur and aitch bone, the top round, the eye of round, the shank, the spider, the bottom round and the knuckle.

Beef Stock

Oh, beef bones. I love them. Rich stock is just one reason why. If you don't know what to do with beef stock, I'm very sorry for you. Here are just a few ideas: braising, marinating, making sausage, drinking warm when ill or when not ill, making sauces, soups, and stews. Cooking grains. Cooking vegetables. Its uses just do not end.

Beef bones, preferably with marrow
Onions, with skins, rough chopped (at the shop we made stock with only onion skins and ends)
Garlic cloves, whole (keep the papers on, even)
Carrots, rough chopped
Celery, rough chopped
Herbs or herb stems
Vegetable trimmings, such as mushroom stems
Water

I don't provide measurements here, because you don't need them. The difference will be the concentration of your stock, so use your sense of smell and make it how you want.

First, roast the bones on a baking sheet, in a 325°F oven until they smell amazing and are nicely golden. Next scoop out some marrow if you like, for making sauces (see bone-marrow horseradish sauce recipe, p. 81), or leave as is.

Place cooled, roasted bones and all other ingredients into a stockpot. Cover with cold water. Bring to a boil, skim off impurities, then simmer as long as you like. You may add tomato paste, wine or other things to flavor the stock. Be creative, and use what you have. Cool, strain with a colander, then store in the fridge, or freeze in batches to use for future cooking projects.

Beef and Lovage Sausage

It's the comfort of beef stew...in a sausage. Lovage is a marvelous plant for your garden, and your kitchen. It grows with almost anything, and many people plant it near root crops. It's a perennial, and can grow taller than a man, so you probably only need one. I adore lovage for its many uses—edible and medicinal. When I was first introduced to it, I was volunteer chef for a fundraiser dinner and concocting a braised beef brisket with surprise aromatics, mostly what could be donated by local growers. Lovage was there, and it paired so perfectly with the beef, I've kept it on my short list ever since. For general sausage-making instruction, refer to the pork section.

32 oz./2 lb. beef lean trim
24 oz./1½ lb. beef fat trim
1 oz. kosher salt
0.3 oz. black pepper
0.6 oz. chopped, fresh lovage
1 oz. cane sugar
0.6 oz. Dijon mustard
1 bay leaf
0.2 oz. dried thyme
0.1 oz. granulated garlic
⅓ cup + 2 Tbsp chilled red wine
10 ft. hog casings

The night before, rinse the casings and leave them soaking in the refrigerator in water. Combine all the ingredients except the meat to make a marinade of sorts. Cube the beef trim (both fat and lean), place it into the marinade, seal and leave overnight.

The next day, remove the meat from the marinade, but reserve the marinade (toss the bay leaf). Flash or open-freeze the trim (place it on a cookie sheet, uncovered, in the freezer until it is almost frozen, about 1 hour). Get your casings ready by fitting the end over your sink faucet and running cold water through them, to rinse them thoroughly and ensure they are free of holes. When the meat is frozen, grind it using a meat grinder, taking care not to force it through. Take half of the ground meat and send it through again. This will create better texture within the sausage, and help everything bind well. Now place the ground meat in a large non-reactive bowl or meat lug and add the remaining marinade. Mix thoroughly. Pinch off some and cook it to test the flavor, adding anything you want to the mix before stuffing. Finally, stuff the sausages and then refrigerate them, uncovered, overnight before cooking.

Braised Beef Shank Tacos with Caper Chimichurri

At my shop, tacos were always the best-selling lunch items. We made them with brisket, steak, smoked pork chop, chicharrones, you name it. But my favorite tacos were always made with shank meat. This braise is an extremely simple, traditional and versatile recipe, so use it with different flavors to create new dishes (some suggestions follow the recipe).

FOR THE BRAISED SHANKS

4 lb. crosscut beef shank, bone in
Sea salt and black pepper, for searing
Some butter
1 sweet onion, chopped
2–3 carrots, coarsely chopped
¼ bunch celery, coarsely chopped
1 bunch fresh thyme
8 cloves garlic, peeled and halved
2 cups white wine
2 quarts water
2 tsp salt
1 tsp black pepper

CAPER CHIMICHURRI

1 bunch fresh parsley (or carrot tops if you're being thrifty)
Scant ½ cup olive oil
⅓ cup red wine vinegar
¼ cup cilantro
3 garlic cloves
1 Tbsp crushed red pepper
½ tsp cumin
½ Tbsp salt
¼ cup capers + 2 Tbsp caper brine

Melt the butter in a large pan. Pat the shanks dry with a cloth, dust with salt and pepper, and sear until golden brown on all sides. Add the onion, carrot, celery, garlic, thyme, salt and pepper, and cook about 5–7 more minutes. Pour in the wine and let it boil, then add the water. Cover the pan with parchment or plastic wrap and then with foil to seal in the moisture. Transfer to 300°F (or lower) oven and cook until the meat is falling off the bone. Remove meat from bones and keep warm while you make the chimichurri.

Combine all the chimichurri ingredients in a food processor and blend until fine. Chimichurri can be stored in the refrigerator in an air-tight container for up to 2 weeks.

To assemble tacos

Warm some soft, corn tortillas on a flat top grill or in the oven. Place shank meat on the bottom and top with caper chimichurri. Add pickled onion (see recipe p. 81), sour cream or queso fresco (see recipe p. 81).

NOTE: This basic braise recipe is very versatile. Instead of tacos, try the pulled shank meat on crostini with anchovy butter (see recipe p. 82) for an appetizer. Or make a sandwich with shank meat and bone marrow horseradish sauce (see recipe p. 81).

Beef Bacon

What a prize. Beef bacon is the best. I don't even fry it up like regular bacon after smoking it. I usually just eat it cold. It's great for breakfast; it's great in salads with creamy vinaigrettes; or it's great just rolled up beautifully on a charcuterie board.

80 oz./5 lb. beef plate or beef brisket, deboned
2.6 oz. kosher salt
1.2 oz. sugar
0.2 oz. cure #1 (optional)

Pat the plate meat dry with a cloth. Thoroughly combine the salt, sugar and salt (if you're using it) and rub generously over the plate meat. Wrap tightly and refrigerate for 1 week. Once the meat has cured, unwrap, rinse and smoke at 180°F until it reaches 145°F. Cool and refrigerate. Enjoy!

Beef Tallow

You've probably used beef tallow before without realizing it. It is an ingredient in many manufactured lotions, candles and cosmetic products. But who needs all that, when you can make it at home? I use tallow every so often for searing beef but mostly, wait for it... I use it for moisturizer. Yep. Just render it, per the instructions below, and then cut it with equal parts olive oil or melted mango butter, then add some essential oil for a fresher scent. You can also use it to make soap, candles, biodiesel, leather conditioner... take it as far as you like.

All you need is suet. Since it is usually pretty crumbled up, you don't need to do a lot to it. However, the more surface area it has the better it will render, so get it into at least half-inch or one-inch chunks. Then put it in your Crock-Pot (or a pot in the oven) on the lowest heat setting (I like to use my Crock-Pot so I can render outside; otherwise my whole house smells like tallow for a while, which is not terrible, just strong-smelling). Stir it every once and again as you go about your day, until it is all melted down and clear-looking. Remove from heat, cool slightly, then pour through a fine sieve or a piece of cheesecloth to strain. Store in the fridge.

Beef Jerky

Beef jerky is a great snack, easy to make, with good keeping quality. The marinade that I've developed is also excellent on a good grilling cut. I've used it on a tri-tip at a party, and the plate was cleaned in seconds. Traditional beef jerky is made with flap meat, but I have made it with eye of round, to make tastier use of this difficult muscle. In Africa, dried and spiced beef is called biltong, and differs from jerky in that it is thicker, uses various muscle cuts, and is never smoked. Prepare either way.

2 lb. eye of round, or flap meat
8–10 dried chipotle peppers, ground into powder
1½ tsp onion powder
2 Tbsp sugar
⅓ cup apple cider vinegar
2 tsp minced garlic
⅓ cup beef stock
2 tsp salt
1 tsp black pepper

Freeze the meat 1–2 hours, then slice it, in the same direction as the muscle grain, into strips. Combine the remaining ingredients into a marinade and refrigerate overnight. Dehydrate 10–12 hours in a food dehydrator and store in an airtight container, or dehydrate in an oven. You can also try smoking the strips for a super traditional jerky.

Bresaola

Bresaola is dry-cured beef that originated in Italy. It is served sliced thin, with bread and olive oil. It is a very easy first dry-cured meat to make. For more information on dry-curing, see Chapter 4: Charcuterie.

Middle muscle in top round, or eye of round
3.5 oz. kosher salt
3.5 oz. sugar
0.2 oz. cure #2
0.35 oz. ground juniper
0.2 oz. dried rosemary
1 oz. ground black pepper

Weigh the meat and record its weight for future reference. Combine all spices in a coffee grinder or spice mill and blend well. Reserve half of the mixed cure, in an airtight container, for later. Rub the remaining cure on the meat and place in an airtight bag in the fridge. Let it cure 1 week, overhauling it (turning it over) every day. After a week, take the meat out and rub the other half of the cure blend on it, then return it to its bag to the fridge to cure for 1 more week. Overhaul it daily.

Remove the meat and wrap it in cheesecloth that is doubled over. Tie the package twice vertically, and a few times horizontally. Now hang it up in a cool place that is not too dry. I have cured bresaola in a springhouse, in my charcuterie chamber, and over my kitchen sink in my north-facing house. Check it daily by smelling it—and I mean really sniff it. Deeply. You'll know if it is going sour. Squeeze it to ensure it is drying. It will be ready after about a month, or when it has lost 30 percent of its original weight.

To serve, slice thinly and drizzle with olive oil.

Sauces and Sundries for Beef

I believe meat shines when it is paired with good companion flavors. Here are a few recipes for sauces and various foods that give beef an extra oomph.

Pickled Red Onion

I almost always have some pickled red onion in my fridge.

½ cup water
¾ cup red wine vinegar
½ tsp sugar
½ tsp kosher salt
⅛ tsp ground cinnamon
1 large red onion, peeled, halved and sliced thin
1 garlic glove, halved
6 black peppercorns
3–4 allspice berries
A few whole cloves

Place the onion, garlic, peppercorns, allspice and cloves into a quart mason jar. Put the water, vinegar, sugar, salt and cinnamon into a saucepan and bring just to a boil. Stir. Pour the hot brine over the onions and spices. Cover and cool. Store in the fridge for at least a night before eating.

Bone Marrow Horseradish Sauce

This is great on roast beef sandwiches, smoked brisket, tacos or appetizers.

¼ cup bone marrow (obtainable by boiling or roasting beef marrow bones and scooping out the marrow)
¼ cup grated horseradish root
½ cup crème fraîche (see recipe p. 150)
1 Tbsp mustard
1 tsp white wine vinegar
Salt and pepper to taste

Combine all ingredients and enjoy.

Queso Fresco

I started making this at our shop for tacos, and now have it at home almost all the time. It is mild, clean and refreshing, providing the perfect balance to the deep flavor of beef.

1 gallon whole milk, preferably raw (but pasteurized will work, too)
2–3 Tbsp sea salt
⅔ cup distilled white vinegar
Cheesecloth, about 2 yards

Stir the salt into the milk in a large stockpot. Bring the milk to 200°F over moderate heat, stirring it now and again as the temperature rises. Don't let it boil. Once you're at temperature, remove milk from the heat and stir in the vinegar. Let the mixture sit for 10 minutes, undisturbed, to let curds form. While you're waiting, line a large colander with the cheesecloth. Finally, strain the cheese through the colander to separate the curds (solids) from the whey (liquid). You should reserve the whey, if possible, for making fermented foods, or feeding to your pigs. The curds will become your cheese. Bundle the curds up into the cheesecloth and strain them as much as you can. Tie the bundle above the sink and let it drip until cool. Then transfer it to the fridge, pressing it between two plates with a weight (like a bottle of wine) on top overnight. The next day, free it from the cheesecloth and you've got queso fresco!

Anchovy Butter

This little recipe is handy for showing off the good marriage of anchovy and cow. Serve it on a beef sandwich or use it with a beefy appetizer along with some zingy radishes.

1 stick unsalted, cultured butter
8 anchovies, packed in oil, drained and minced
Half of a shallot, peeled and minced
½ tsp lemon juice
½ tsp dried thyme
Sea salt to taste

Soften the butter (do not melt it), then combine it with the other ingredients in a food processor. Form the mixture into a log or brick, wrap in parchment paper, and refrigerate until firm.

CHAPTER 2

Lamb

As I enter the barn, the light is filtering through perfectly, and I can hear faint bleating from the stall. We're slaughtering three lambs today. I let myself into the pen. The oldest, Hercules, is standing on the ramp. The smell of hay and fleece and dung stamps a memory of him standing there, with his head turned to see me, his eyes shining in the half-light of the morning. I drove all the way here with my troubles thick upon my back. I met a friend and borrowed a tool. He said to me, "You're shaking. You're a wreck." He put his hand on my forehead, like a mother to a child. "You're gonna be alright," he said, his eyes upon me with an intensity of concern and a depth of reassurance that struck home.

Now, standing on the hay-covered ramp, looking into the eyes of this animal we will kill, I see the same conviction, the same knowing. Not about me, but about life. About what is. I find tremendous humility in those eyes. I snap a photo of Hercules, there in the dark barn. I vow to get over myself. I feel my body, alive. I promise to keep it steady.

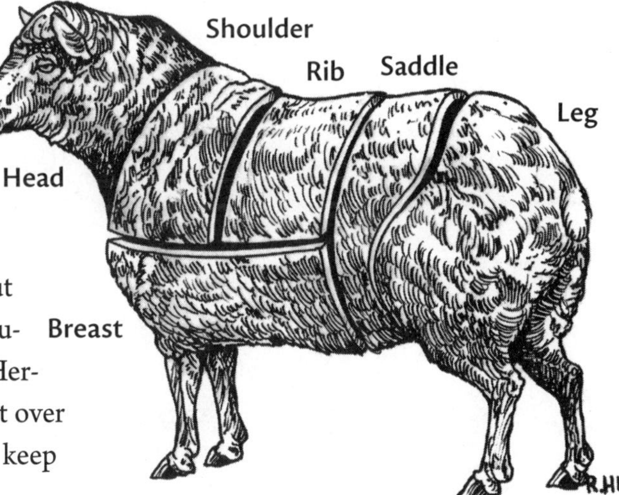

When everything is set, Tim takes his lamb into the grass. He picks Hercules up, and puts him back down on his side, in the pasture. With gentle but firm hands, he pulls the animal's ears down over his eyes and begins to stroke Hercules' shoulder and neck. He is leaning in close, talking. I can't hear what Tim is saying but his voice is calm, soothing. A few times Hercules starts, his head popping up to look about him, but Tim reacts little, calmly replacing the ears and starting over, all the time talking to Hercules. When the lamb softens and is still, Tim reaches back for the knife. In a confident, bold stroke, he cuts Hercules' throat.

In my work, I have watched many animals pass. In my classes, people ask me sometimes how to ensure that it will go well. I've even had people ask me if there is a way to make it easier. Of course, the discussion flows into humane slaughter regulations and stories of individual experiences with various methods. But I am always sure to mention one thing: It is never easy. It isn't supposed to be.

Now, as we stand or squat in silence behind Tim, watching Hercules' body tremble as his nerves send final shockwaves through his muscles, the mountains rise up in front of us. The incredible morning mist hangs just above, and stacked atop it, the clearing sky. This silence, here, is cradled by all of these things. I won't even shift a boot in the grass. I will not take my eyes off of Hercules. Even after his body lies still, we wait, until Tim slowly nods his head and rises. We lift Hercules to the gambrel.

As the day wears on, we work hard. We learn from each other. We discuss the best uses of all the parts of the body. We separate hearts and heads, we roll fleeces for later washing and tanning. We slaughter two more lambs, Maybelle and Whiteface. We share silence and sadness each time. Each time, I feel my breathing, and I feel gratitude, for precious life as well as for suffering.

Inside the space of silence, I think, *I could stand to die here*. Here, in this pasture, under these mountains, this sky. If, when my time comes, someone lays me down in green grass and soothes my body, and someone I love then lovingly cuts my throat, and my blood runs into the grass and fertilizes the soil, and my body is used for food and necessity, I could handle that. If five beings, alive to the world, come together over my passing, and learn and cooperate, and even turn my stomach inside out to marvel at the mystery and beauty of nature, well, I think that is one of the best deaths I can truly imagine, much more preferable than what many humans can hope for.

Recently, my days have been like people, the way they rub me. I know almost as soon as I meet each one. The past few have been an impatient crowd; times are hard, but today has been much more in focus. Is this a gift from Hercules and the other lambs? Something more than meat and bone? Something softly bleated, barely heard in the boom of everyday life? Is it because I am letting the baby sleep with me tonight, his tiny arms up above his head, his whispering hairs, his uneven breathing keeping me awake? I lie in bed, thinking about Hercules and the sheep at work, moving like a hushing river through the grass. I love them. I love their many heads all turning at once.

Raising Lamb

Lamb is, in many ways, forgotten meat, and sheep are a forgotten resource in America. The sheep industry saw its peak in the late 40s, but herds have declined almost ten times since then, with only about six million head of sheep raised annually in the United States. The major sheep-producing states are concentrated in the West, and most of the US stock is concentrated into large flocks of more than a thousand head. For perspective, consider that 91 million head of cattle are raised in the United States annually. That's fifteen times more beef than lamb, which means Americans consume less than a pound of lamb each, every year. Per capita consumption of chicken, on the other hand, is close to a hundred pounds.

As one might imagine, this makes the economics of lamb even more backward than meat economics already are. As with most meat, on the industrial scale, lamb is more expensive to produce than it is worth. Lamb also has the unfortunate and unique distinction of being more expensive to produce than the average American can afford. So it's almost astonishing that we still raise sheep at all in this country. It is seen as a niche market on almost any scale, and really catering to the seasonal gourmet market and the ethnic market. Exports cover a small percentage as well. Meat is the name of the game with sheep production, for the most part. Wool is an incredibly small market in the US, most of it exported, and on the larger-scale farms, wool has virtually no value.

What sheep we do raise, we handle like our other animals. The typical American lamb is born on pasture and stays by its mother's side until about eight weeks of age, when it is weaned and transferred. Many lambs are *backgrounded* at this stage, meaning that they are transferred to another farm, where they graze some more before being sent to the feedlot. Once there, the animals are raised with high stocking densities, antibiotics and grain rations for flavoring and fattening. Slaughter comes as early as 90 pounds for the ethnic and lighter-weight markets, but lambs may also be taken to as much as 150 pounds for the heavyweight market. Depending on breed, lambs can finish at under one year of age.

Lamb, like beef, can be raised strictly on pasture, but grain feeding is customary in the United States. Some producers *creep feed* their lambs, meaning they start introducing a controlled, small amount of grain just after weaning. This practice has been found to increase growth, as younger lambs are more efficient feeders, converting food to meat more easily than larger animals. Grain-fed lamb is more appealing to the American palette, as we are not used to the strong flavor of 100-percent grass-fed lamb. However, concern about antibiotic resistance, genetically modified organisms and animal welfare drives the grass-fed lamb market, mostly of imported meat.

The entire lamb conundrum seems to stem from cultural bias, as well as the regulatory history within the sheep industry. I'd also wager that American land managers and consumers alike know less about lamb than we'd need to build a booming lamb economy, simply because the market has not incentivized an effort to develop the most consistent product. Add to that the constant influx of lamb from New Zealand and Australia, which makes up half of the US lamb supply. Because this imported lamb is strictly grass-fed and of smaller frame than typical American lamb, many people have had variable and inconsistent experiences both cooking with and eating lamb.

Finally, the age of an animal is all-important when it comes to sheep meat. Hopefully, you are beginning to learn that the manufacture of quality meat is a complex process and relies on the intricate workings of nature, not just manhandling. For example, we've discussed how muscles change as they mature, how muscles taste based on how the animal uses them, how breed effects flavor, how the presence of fat increases flavor, aids cooking and facilitates storage, and how feed and forage changes muscle tone and color. Ask yourself, if a lamb is given eight months to eat and live, how should you expect its muscles to taste? Green as grass, young and elastic, bright and wild. If America loves deep flavor (which it does), why are we exclusively producing such young meat? And why is the sparse market for this amazing animal mostly concentrated in the spring, a season that is not favored by that animal's natural life cycle?

I'm making a case here for mutton: sheep's meat from older animals. Mutton is not from worn-out, culled sheep, as the mainstream would have you understand; it is simply meat from older animals. How much older? True lamb meat comes from an animal slaughtered at 12 months or younger. Mutton can refer to meat from any animal older than this, although abroad, there are different terms applied to a range of ages. In Europe, *hogget* is a term used to refer to 2–3 year olds, and *mutton* describes meat from animals 3 years and older.

Regardless of when we decide to call our lamb meat mutton, the point is this: Why don't we raise honest, grass-fed lamb like we raise honest, grass-fed beef? What if we took animals to 18 months, or even 2 or 3 years of age, allowing them to digest plenty of diverse forage, use their muscles, and develop lovely, flavorful fat stores, under natural conditions? At this age, with this fat, we could hang and age the carcasses post-slaughter, just like we do beef. We'd be eating great meat. Perhaps we'd grow a steadier, more consistent market for the product as well.

It goes without saying that lamb is under-used and underestimated. As we seek alternatives to the status quo, we could stand to gain a lot from this animal, including food, fiber and valuable management of our changing landscapes. This section will cover introductory information and considerations for production of lamb on the homestead. That said, there is much to learn, and I'd encourage further reading as well as experimentation on your own land holding to develop optimum holistic synergy; produce consistent, high-quality meat; and bring lamb into the light it deserves. Several resources are recommended in the Resources section.

Breeds

Genetics may be one of the most daunting, yet most amazing, points for consideration in sheep production. Because sheep have so many uses, there have arisen many unique breeds. These breeds are categorized by purpose first, into wool breeds, meat breeds and multipurpose breeds. Further classification within these categories has to do with specific characteristics, such as fine wool vs. long wool, blackface vs. lightface, short or long tail and more. Sources can't seem to agree on how many distinct sheep breeds exist worldwide, but I've heard it recorded at two hundred, and also seen it claimed at a thousand.

As a homesteader, this matters less, but also makes it harder to decide. The best way to begin is by considering your purpose. Of course, the object of this book is to identify breeds for quality meat production, but from the homesteader's perspective, the option of combining meat breeds with dairy animals, or wool breeds with meat breeds, is pretty enticing. And if you're interested in exploring mutton, much of the existing information about the meat quality of specific breeds focuses on lamb; the details of those breeds' meat at older ages may not be known.

As with any animal production, tailor the genetics of your herd to what you'd like to accomplish, but with sheep, consider climate doubly important. Because of their wide genetics, sheep vary greatly from breed to breed in their parasite resistance, their prolificy (ability to birth large litters) and their ability to feed efficiently in extreme weather.

Common meat breeds are the hair breeds Katahdin and Dorper and the wool breeds Dorset, Oxford, Hampshire, Suffolk and Southdown. Depending on their origin, these breeds will vary greatly in parasite resistance and overall frame. Choose breeds with some hair sheep lineage if you do not intend to shear. Hair sheep are different from wool sheep in that they do not require shearing but will shed their thick coats instead. Also consider biosecurity, to avoid purchasing animals that will bring diseases or parasites onto your farm.

My experience in raising sheep has been limited to hair meat breeds crossed with wool meat breeds and hair meat breeds crossed with rare breeds. Each has had its benefits. I live in the Southeast, in what is essentially a temperate rainforest, so high parasite resistance is desired, along with the traits everyone wants: heavy lambing and good muscle tone. As such, producers in my area are crossing standard meat breeds, of course, but also hair meat breeds and tropical breeds that have evolved in wet regions with many parasites. Recently I've been working with a herd of St. Croix–Gulf Coast Native crosses. Most others in my area are working with a Katahdin–Dorper cross.

Some thoughts for consideration, from both research and experience:

- Hair sheep and tropical breeds are pre-adapted for warm, wet climates.
- Long-wool breeds are pre-adapted for cold, wet climates.
- Fine-wool breeds fare best in dry, hot climates.
- Prolific breeds and dairy breeds are the most adaptable, because they have evolved in many, varied climates.
- Don't worry about purchasing specifically for prolificy. A ewe with only 25 percent prolific ancestry can produce litters of the same percentage as a purebred prolific ewe.

- Consider ram size carefully when breeding, and ask about the sire (father) of purchased stock, as the characteristics of the ram effect the meat quality of its offspring. Seek large-frame rams if you intend to feed your lambs to 100–140 pounds, and small-frame rams for a finished weight of 60–90 pounds. If you're raising mutton, this may not matter as much, but erring on the side of larger sires would seem wise.
- It seems that small-frame sires create better pasture adaptability for lambs. This is unfortunate for meat producers with the intention toward the most honest product. A solution might be creative cross-breeding of sires to create reliable, grass-fed, heavyweight lamb or mutton.

Space and Water

Plan for 25 square feet of living space per animal. If the flock is moved daily, forage is abundant and animals are weaned, you can stock at a higher density. Ewes with nursing lambs need the most space, so if your herd is mixed, planning for 25 square feet per animal will allow for nursing ewes, rams during breeding and all lambs.

Parasites are best controlled by good stocking practices. Remember that parasites live in animals but complete portions of their life cycle in the soil and manure droppings. This is the best argument for conscious and well-managed rotational grazing. Making sure there is at least 60 days of rest before a paddock is re-grazed by lambs or goats will assist in parasite control. Where grass is taller, parasite contamination is less of a threat, as the parasites reside very close to the ground. As such, overgrazing will increase parasitism in the flock.

Lambs can withstand varied climate situations, and can even graze through considerable snow. Ventilation and dryness are more critical than temperature. Naturally, variation in temperature and moisture tolerance occurs from breed to breed.

Water is one of the most important considerations when raising any animal. Responsible production requires fresh, clean water for all animals without depleting natural water resources. Do not let your animals drink from the creek. Their manure in the creek is a disturbance to the natural aquatic ecosystem, and can introduce harmful bacteria that will travel with the water. Fencing the animals out of nearby water bodies, but tapping into these water resources via piped irrigation or pumps, is a cost and effort that you will have to assume. See the Resources section for further reading on setting up watering systems and equipment to make watering and rotating more efficient.

Fencing

At the farm where I currently work, we run cattle, sheep and pigs in the same pastures, so we have highly adaptable fencing. We have a high-tensile perimeter fence and run two-strand poly-wire cross-fencing to create the grazing paddocks. Sheep and cattle, as long as they have a good selection of greens to eat, rarely challenge the fences. Another option for sheep, and really the best option for goats, is netted fencing, either electrified or made of durable wire. You can source nets from many suppliers, and affix them to step-in fiberglass posts. You may choose to fashion corner and end posts from wood, giving the fence more durability.

Predation is an issue with small ruminants such as sheep and goats. Guard animals such as donkeys or certain breeds of dogs are often kept with small ruminant flocks for protection. Wild predators of larger flocks in the Midwestern United States have been the cause of much controversy, as government-funded agencies are charged with killing wolves, coyotes, wild cats and other predators that threaten domestic animals. Over time, we have learned that disrupting the predator populations has drastic and negative effects on ecosystem balance, and in some cases (as with coyotes) actually triggers a rise in population. Guard animals and good fencing are the most benign ways to protect your flock. If you do need to take other measures, consider the ecosystem costs. In nature, the larger the flock, the less effective predators are at removing individuals. Owners of small flocks should consider paddock placement on the property, and move animals frequently as additional protection.

Sheep prefer some browse (woody plant foods) and appreciate shade in their paddocks, so be careful about electric fences becoming overwhelmed by brush and timber. Keep fence lines clean with machetes or weed eaters on a regular basis.

Feed and Minerals

Sheep can graze mixed-grass pastures for general maintenance but also prefer some percentage of forbs (broadleaf plants such as chicory, often called "weeds"), legumes and woody browse. This offers plenty of opportunity for creative rotation. Sheep can run in paddocks of half pasture and half timber, or in spent crop fields. Along state highways around Asheville, people are putting goats and sheep to work eating kudzu, poison ivy and more. Here's a breakdown of how sheep and goats prioritize plant foods to make up their ideal diet:

	Sheep	Goats
Grasses & Legumes	60%	20%
Forbs	30%	20%
Browse	10%	60%

You can see from the table that if you're looking to manage mostly pastures or croplands, sheep are your best bet. If you have woody plants to manage, for the most part, goats are better companions for your land. Goats will girdle young trees (creating a gap in the bark so nutrients cannot travel continuously up the trunk), so be conscious about where you allow them to browse freely. You can provide some measure of protection for trees you are cultivating, or deem especially important in the landscape, with guards or fences around the trunks.

You'll also want to consider the composition of the small ruminant's diet based on the demographics of your flock. Pregnant and lactating ewes will require more calcium, which is best delivered by legume forages. Many producers supplement with legume hay (such as alfalfa) during gestation, or feed a grass/legume mix. The stage of growth at which hay is harvested is the biggest determination of the hay's nutritive quality, so make sure you know where good hay comes from in your community. All grasses and forbs are more nutritive when they are in their vegetative state (before they set flower or seed), so manage grazing accordingly, and note the presence of seed heads in hay bales.

If you choose to supplement some or all of your lambs with grain, use care in choosing and introducing the supplement. Many large-scale producers begin feeder lambs on creep rations of corn and corn by-products early in their life to allow for the most efficient feed conversion and to allow the animals' rumens to adjust. Sheep are more sensitive to introductions of new feedstuffs, and random overload of a new food source can cause internal toxicity or imbalances in the rumen. Choose top-quality grains (either organic or grown on your land) and introduce them slowly.

Sheep and goats will both require minerals. Free choice rations of minerals will allow the animals to select what they need. Many providers can supply mineral mixes, and I have made recommendations in the Resources section.

In my mind, sheep and goats are wonderfully versatile. When they can be used to manage so many different types of landscapes, I don't see a reason to supplement with grain except to affect flavor. Coupled with the implications and costs of sourcing grain for animal feed, this makes supplementation seem questionable on the home scale.

Additionally, if invasive plants (see sidebar) are palatable to these animals, and are out of control by your standards, by all means, make use of the animals. I have a huge argument mounting for our ability to cooperate with our meat to make a more beautiful and nourished world. Livestock animals present us with an opportunity to manage landscapes with a style that makes more sense to our earth. If a particular plant species is problematic to you, know that you may take many steps to eradicate it, but you will not eradicate all your problems. Consider a more natural approach. Your goats or sheep will also effect changes that the ecosystem will better take in stride. If kudzu is encroaching, let your sheep or goats graze it. They will manure the area;

their hooves will scarify seeds underfoot; and the action of their eating will change hormonal growth responses in other plants. These and other ripples will slowly and miraculously paint a larger, more diverse and more colorful canvas on your land holding.

When I own land again, I'll be working on a manifesto for multi-species grazing, and it will most surely include sheep. Creative rotations between pasture, wooded areas, carefully selected cover crops and crop stubble will allow me enough removal time to control parasitism within my herd, will provide the animals with a diverse and flavorful diet (read: delicious mutton) and will help me manage my slice of an awesome ecosystem with less avarice.

Invasive Plants

Invasive is a term used to describe plants that are growing outside of their native habitat and taking over. In the Southeast, a perfect example is kudzu. The general practice for dealing with these plants is to spray them with pesticides or dig them up at the root. What is missing from the conversation about these plants is their role as environmental indicators, or reminders of where we might analyze our impact on that ecosystem. What have we done to allow these plants to establish? Beyond introducing their seed or stock, chances are that people have wrought some major disruptions to soil, water or diversity. Invasive plants are calling cards of a natural world that is more complex than we can understand. This is not to say they are not problematic. Indeed, kudzu is taking over on roadsides, limiting biodiversity both above and below the soil. Chemically motivated eradication programs are the silliest response we can possibly have to such a message. I guarantee that the so-called problem plants are not going anywhere, at least not as a result of our current interventions. If they do, they are likely to be replaced with other plants that surprise us, plants that we'll again call "problems."

The way ecosystems evolve is through a process called *succession*, wherein species that can habituate and thrive will establish first, thus changing the game before they make way for new species to enter the scene. The complex, working relationships between

Lamb Butchery

Lamb is easy to work with because it's small. You won't want to worry about splitting it in half until after you've cut it into primals (unless you want to smoke intact halves, which is a very worthy endeavor indeed). If you do want to split the carcass in half before proceeding, I have used a circular saw, as well as a reciprocating saw, while the animal was hanging.

The lamb carcass comprises four primals: the shoulder, the rib, the saddle and the leg.

I visited my good friend Matt Helms at the Chop Shop in Asheville, NC, for our lamb breakdown. The lamb pictured here dressed out at

all species is truly beyond our powers of imagination. Succession is clearly visible in landscapes that have endured some disaster, such as a fire. You can also easily observe it happening in urban environments. If you shift perspective, you can start to see that a parking lot is a huge cataclysm for the natural communities that once thrived in its unpaved, untouched space. The lichens, mosses and grasses that you see establishing themselves in the cracks and gouges of the asphalt are tangible signs of succession, a brilliant recovery plan. Of course, humans are (too) well established here, and the parking lot won't just return to its original state with us driving and walking all over it. The natural notes you barely notice in the lot will most likely stay sort of "stuck" in a certain stage of succession, due to our repeated interruptions. But nature does not stop trying.

Now imagine all the ways we have messed with succession: by planting crops, dropping bubble gum, digging holes or merely breathing. It's mind-boggling. As we go, we learn that many ecosystems evolved with animals as a major part of their natural succession. Some seeds need scraping or smashing before they will grow (read: hooves), some need extra fertilizer (read: poop) and so on. Knowing that by merely moving, we are affecting succession, why not try to affect it in creative ways that promote healthy diversity? Alan Savory says, "Managing plants and animals in isolation is meaningless." I couldn't agree more. ◆

Our pasture-raised lamb.

48 pounds, and was raised on pasture with a free choice ration of grain throughout its life. As such, the meat is much paler than that of a strictly grass-fed lamb. I typically work with grass-fed only and was intrigued by the difference in muscle development and color due to grain supplementation.

First, Matt removes the neck. He starts with the knife, right where the spine begins to curve into the back, cutting all the way around the bone and then finishing with the saw. The neck is excellent roasted or braised with the bone in, boned out and tied before roasting, or deboned and cut into chunks for stew. You can also debone it and contribute its meat to the grind, of course.

Next, Matt begins the work of isolating primals. Rather than starting from the front, he prefers to work from the back. This makes the

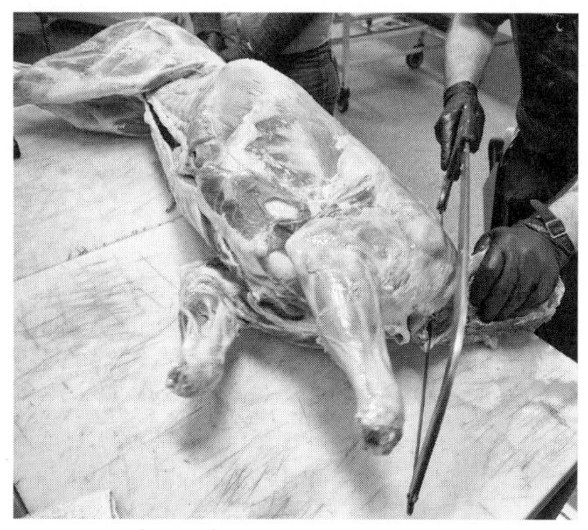

Removing the neck.

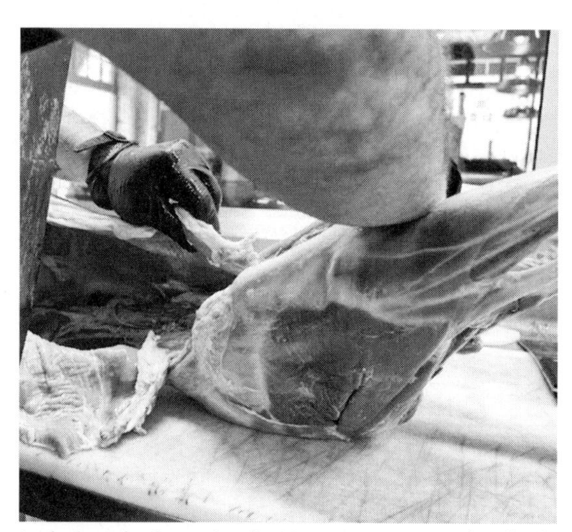

Isolating the leg primals. Here you can see the flank that Matt has cut away from the leg, following the outer profile of the leg muscle itself.

carcass easier to handle on the bench. He will remove the leg primals first.

Start by cutting the flank meat (part of the saddle section) from the legs. Using your boning knife, follow the outline of the leg muscles, and follow that angled seam until you reach the muscles in the back.

To fully remove the leg primals from the carcass, use your saw to sever the spine about two vertebrae up from the tail. You want to separate the leg primals just where the tenderloin ends.

Next, Matt isolates the saddle primal, which contains the strip loin and flank meat. He makes his cut between the 12th and 13th ribs.

Finally, we need to separate the shoulder primal from the rib. Do so by making a cut with your boning knife between the 5th and 6th ribs, and sawing through the spine and breastplate.

You can see our work thus far in the photo below.

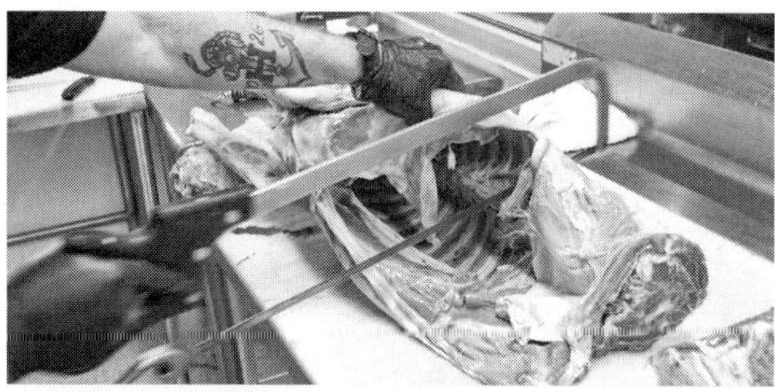

Separating the saddle primal.

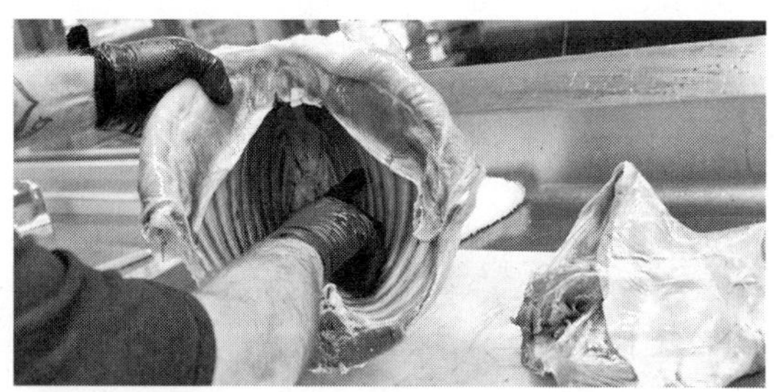

Matt counts ribs.

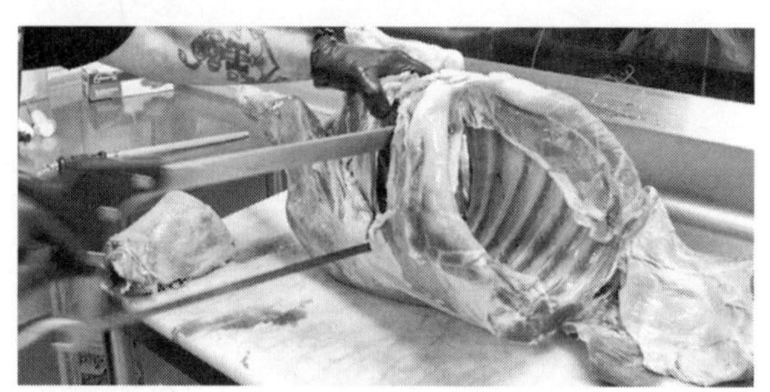

Separating the ribs from the shoulders.

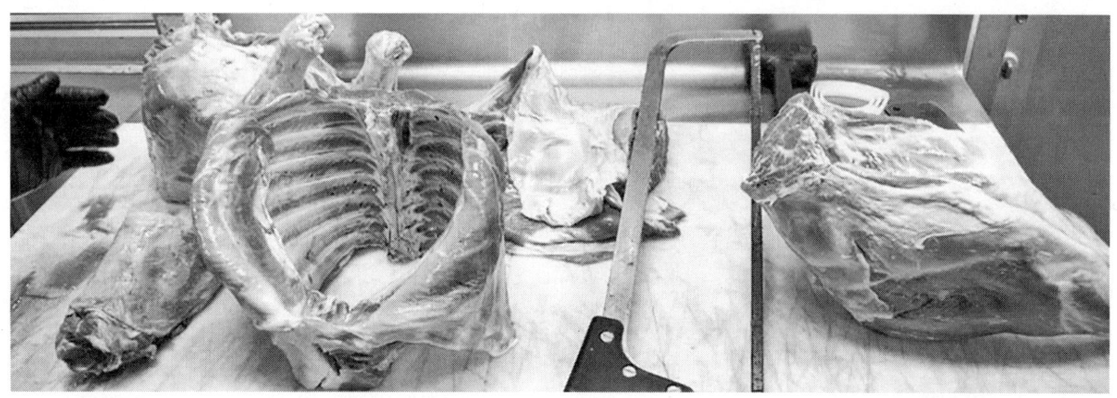

From top, left to right: The shoulder portion, the neck, the saddle, the rib section and the legs.

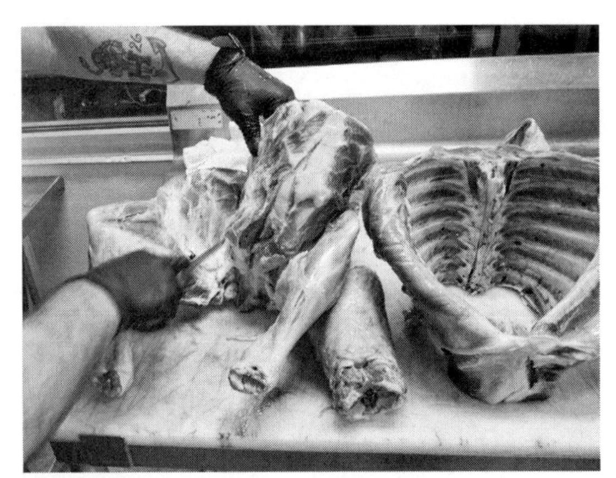
Matt removes an arm.

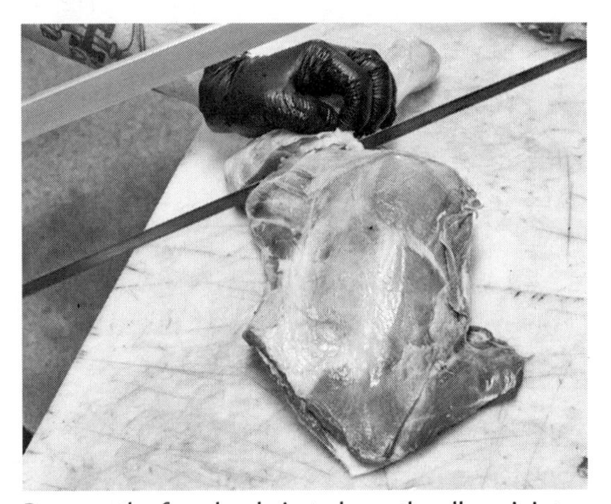
Remove the foreshank, just above the elbow joint.

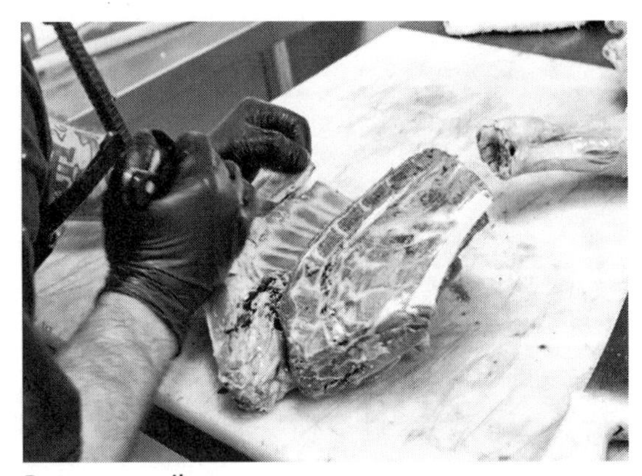
Remove cartilage.

The Shoulder

We'll continue with the shoulder primal. Refrigerate the other primals until you are ready to work with them.

We'll demonstrate cutting chops with one side of the shoulder carcass and a bone-in shoulder roast with the other side.

Since lambs don't have a lot of tissue connecting their breasts and upper arms, you'll find there is a lot of movement in the shoulder. Blade chops are traditionally cut through the entire piece, but this can be harder to do when working with a hand saw, since you'll have to hold the arm steady while you saw. Instead, you can cut chops from the breast and shoulder separately, by first removing the whole arm. This is an easy and intuitive cut to make.

Next, remove the foreshank from the arm. You can braise the shank bone-in, or bone it out and contribute its meat to the grind pile.

Set the shank under refrigeration. Now flip the breast over. Remove the cartilage and feather bones from the tip of the breast by cutting with your boning knife along the end of the rib bones.

You can stop here, and what you'll have are lamb Denver ribs. They can be prepared like you would spare ribs. Many people stuff the breast and roll it for roasting. Since it has decent fat, you can use it to make lamb rilletes or other lamb charcuterie. (See the charcuterie chapter and adapt the duck rilletes recipe for your lamb breast.) Another option for the breast is to take it a step further and cut chops,

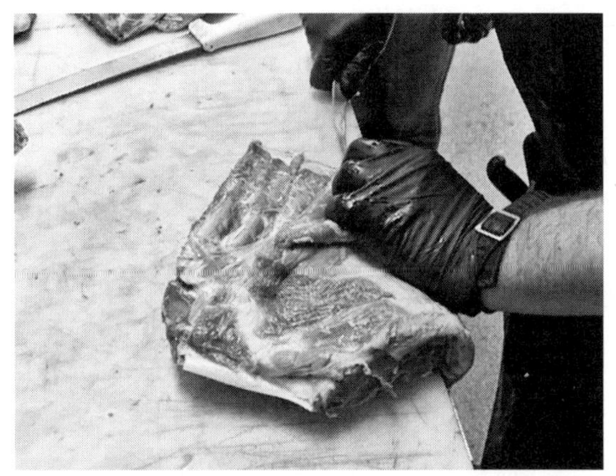

Cleaning plate meat from the ribs of the breast.

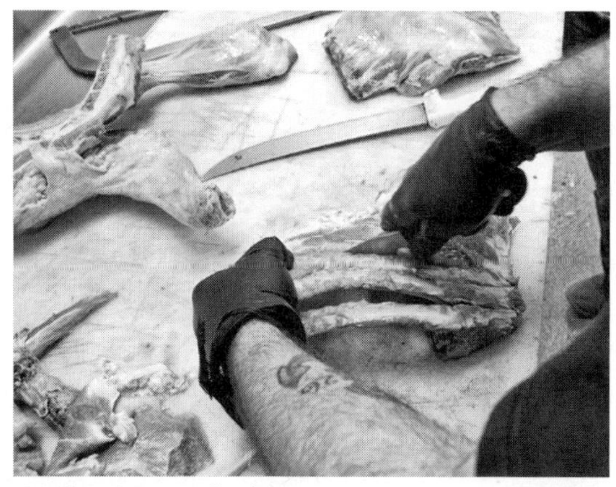

Cutting Denver rib chops.

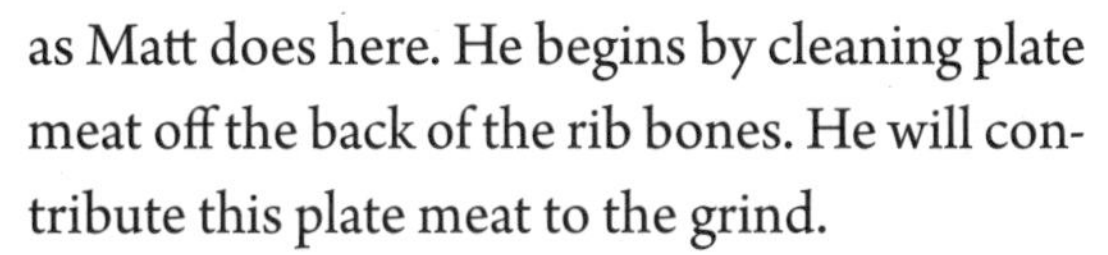

as Matt does here. He begins by cleaning plate meat off the back of the rib bones. He will contribute this plate meat to the grind.

Next, use your boning knife to cut between each rib and mark off the chops. You'll finish with your cleaver, to break each chop cleanly from the breast. In this case, Matt will call these lamb shoulder chops, since they are from the shoulder section. You could also call them Denver rib chops.

You should have a shoulder remaining, from which you've already removed the foreshank. In our example here, we simply crosscut the shoulder to produce more chops. You could do the same. Alternatively, bone out the shoulder and use the meat for sausage, smoke or roast it whole, or divide it into smaller roasts. The shoulder is also great for smoking on the bone.

For the other whole shoulder section, we're simply removing the shank and keeping the shoulder and breast intact for smoking.

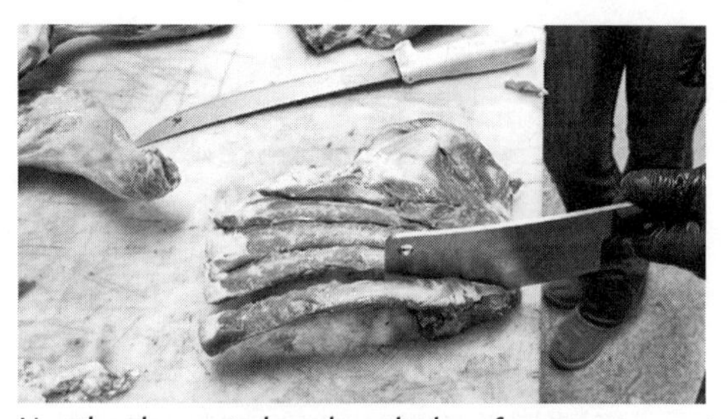

Use the cleaver to knock each chop free.

Lamb shoulder cuts. From left to right: trim from the breast and bones for stock, the other whole shoulder and arm, foreshank, shoulder arm chops, denver rib chops (from the breast).

Removing glands and tissue from the cut shoulder.

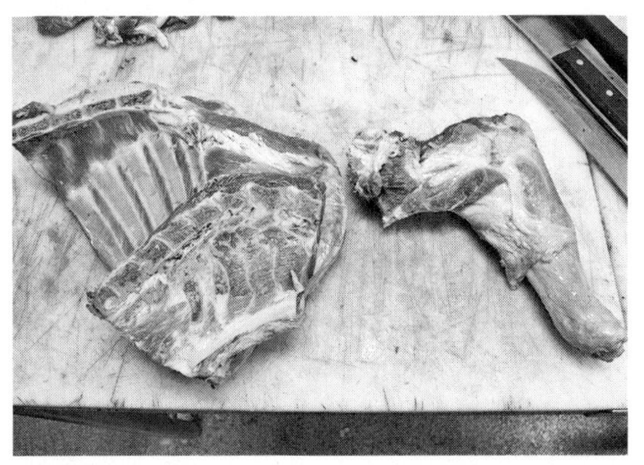

Shoulder and breast bone-in, foreshank.

Remove the shank as before, just above the elbow joint. Then flip the remaining shoulder and breast over, and clean glands and tissue away.

You can trim up the shoulder cut as much as you want, removing rib tips and featherbones, messing with the chine—or both or neither. I'm leaving it intact for a smoking project. The photo below shows the result.

The Rib

Next we'll deal with the rib section. Start with the ribs upended, so you can use your bone saw to easily split down the middle of the spine.

Next, square off the prime rib. Find the tip of the 6th rib (shortest on the rack) and mark a line straight across from its tip. Use your saw to get through the rib bones, and finish with your boning knife. You've removed the lower "spare ribs" and the skirt.

Next, turn the rack on its end and saw

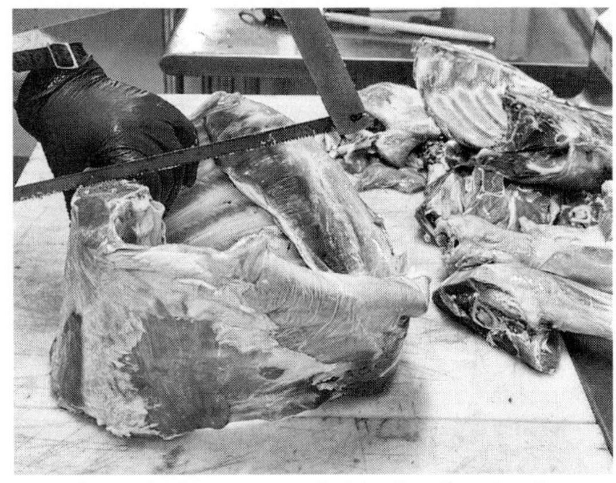

Saw through the spine to divide the rib primals.

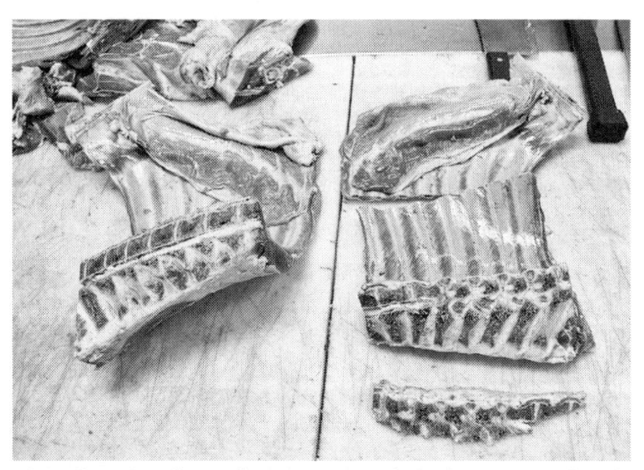

The rib primals. Left: Intact. Right: The spare rib, rib rack and chine bone, separated.

the chine (spine) off. Note: If you are going to leave the rack whole for roasting, skip this step.

Save the chine for the stockpot. You can, of course, smoke or grill the spare ribs, but you'll find an alternative recipe on page 114.

Next we'll french the rack. First, pull the plate meat off the outside of the ribs and contribute that meat to the grind.

Flip the rack back over, so you're looking at the inside. With your boning knife, firmly cut a line down the center of each rib bone, through the tissue, two to three inches long.

Next, find the halfway point between the loin and the rib ends. Make cuts parallel to the loin in between each rib bone, to denote where you will stop frenching the bones.

Ready your patience, and put down your knife. Using your fingers, peel back the tissue atop the 6th rib (first rib on the rack), beginning with your mark in the center. You'll need

Another angle on the rib primals. Top: intact. Bottom: The result after removing the spare ribs and the chine.

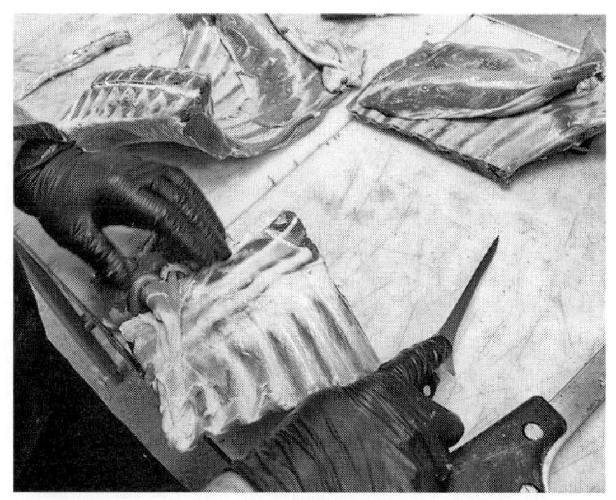

Matt removes the plate meat from the outside of the rack.

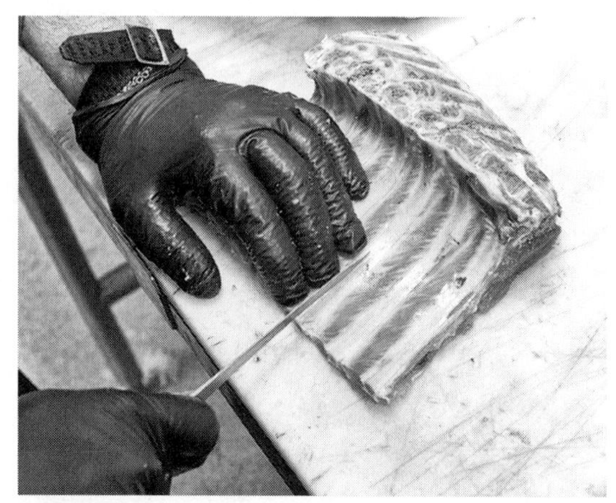

Cut down the center of each rib. You'll come back to this later.

Make cuts in between each rib, about halfway to the loin, to mark where you will stop frenching the bones.

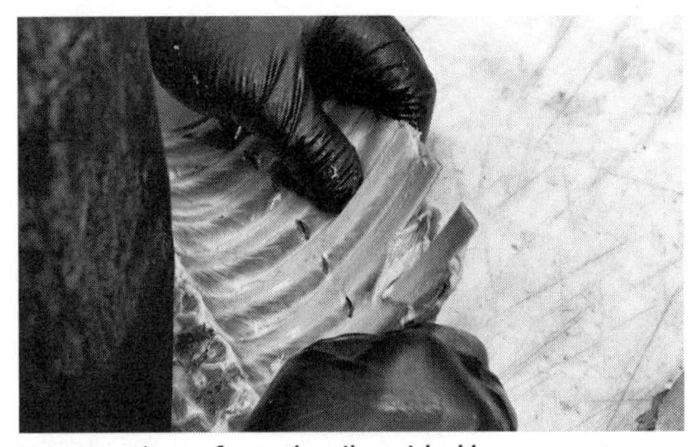

Remove tissue from the ribs, aided by your cut down the center inside of each bone.

to pull the tissue completely away from the ribs one at a time, in order.

As you go, curl your index finger around each rib, and pull downward to remove the meat and tissue from each bone.

When you've reached the end, you'll have the rib meat hanging from the outside of the rack. Remove it in a clean cut with your boning knife. Throw it in a pan and have a snack, or put it in the grind.

If you are going to roast the frenched rack whole, you'll want to leave the chine bone on, so that the loin is not resting on the roasting pan during cooking. In this example, you'll recall that Matt removed the chine first thing. This was just so we could cut frenched rib chops.

To cut the rib chops, start by marking each chop with your cimeter, in between the rib bones.

Then knock off each chop with your mallet and cleaver.

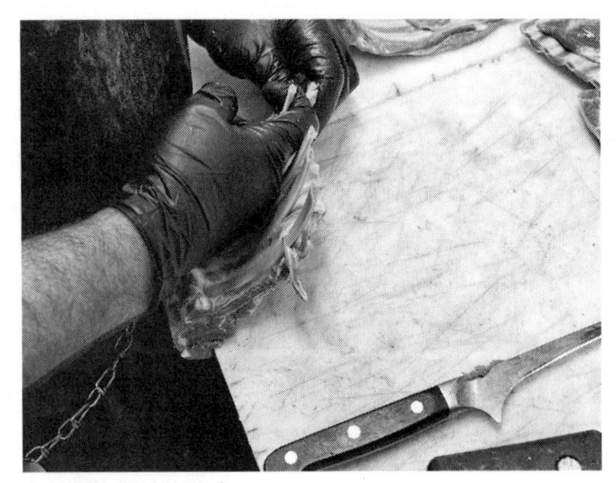

Frenching the rack.

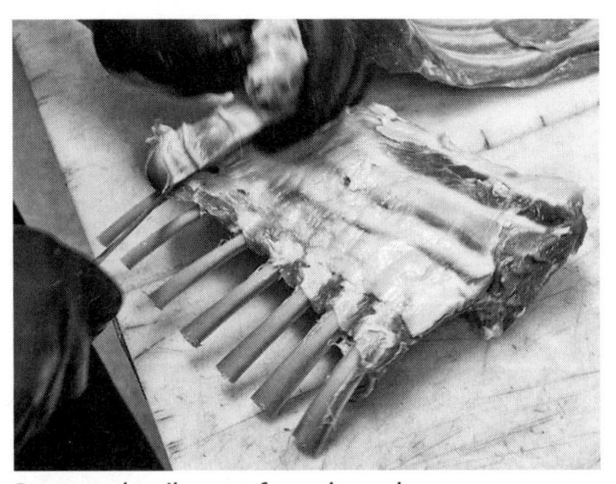

Remove the rib meat from the rack.

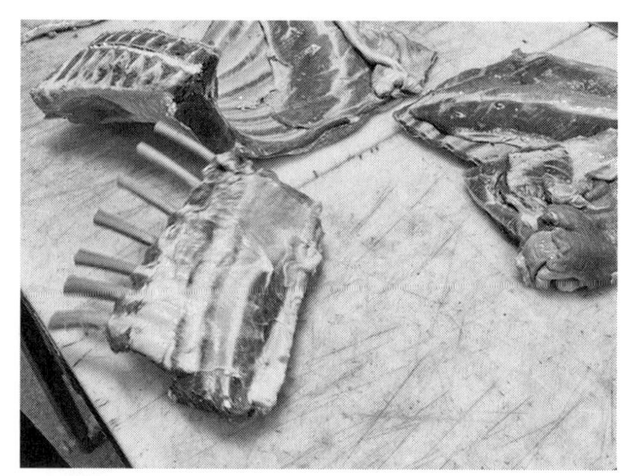

The frenched rack.

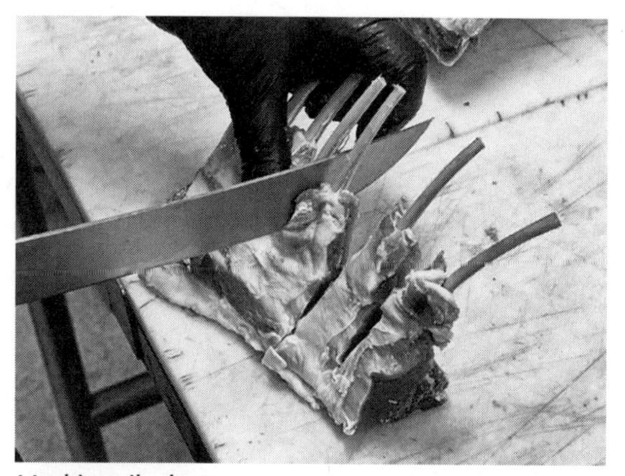

Marking rib chops.

Cook your rib chops on high, dry heat, taking care not too overcook. Try preparing with a compound butter, with lemon and paprika or a simple herb and garlic seasoning.

The other rib primal, which you've left relatively untouched, is suitable for roasting whole. Remove the skirt and rib cartilage at the rib tips (contribute these to the grind pile for sausage) to create a tidier square cut. Roast according to the recipe on page 113.

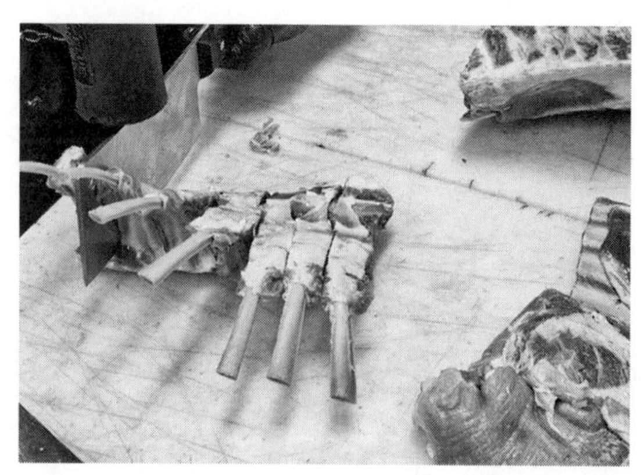

Using the cleaver to break off each rib chop.

The Saddle

Next, we'll work with the saddle, which includes the strip loin, tenderloin and flank. Begin by splitting down the center of the spine with your bone saw.

Remove the 13th rib and add it to your stockpot.

My favorite thing to do with the saddle is lambchetta, which is loin wrapped in belly. Start by boning out the loin. Cut close to the bone, leaving as much meat on the loin as possible.

The finished, frenched rib chops.

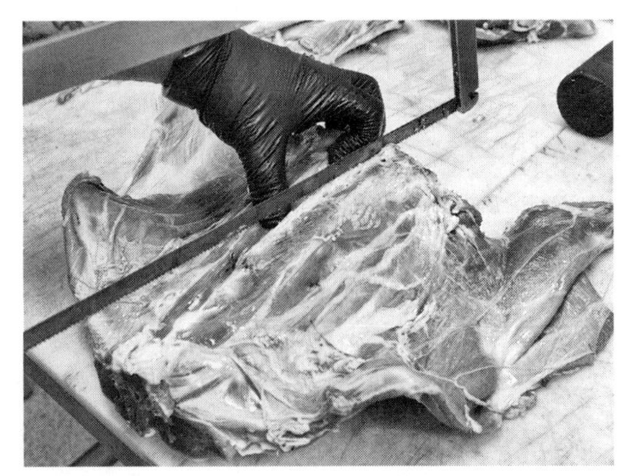
Splitting the saddle section.

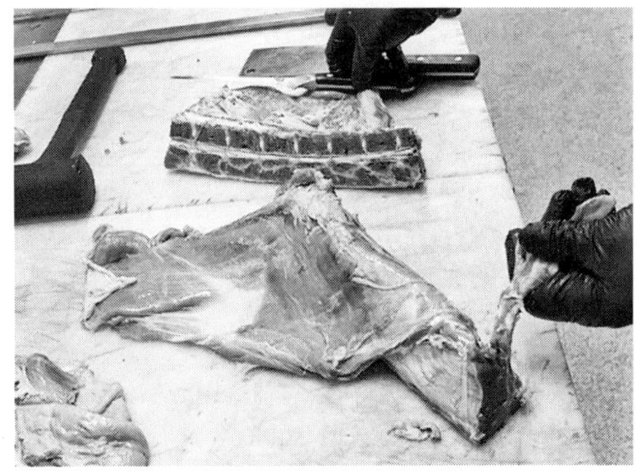
The saddle primal with the 13th rib.

The tenderloin will remain on the bone, unless you've deboned with enough finesse to remove it along with the strip loin. I like to include it in the lambchetta. If it's still on the bone, take some time to carefully remove it now.

Turn back to your developing lambchetta and nestle the tenderloin up against the strip loin, right in the seam between the flank and the strip loin. Next, starting with the loin, roll the loin and tenderloin into the belly, like rolling a carpet. Try to keep the piece as even as possible.

If you're preparing my firecooked lambchetta (p. 112), you won't want to tie this roast off until much later (once it is seasoned and wrapped in cloth). In general, you want to season a roast before tying it anyway. In this case, Matt will go ahead and tie it off, placing butcher's twine every two inches or so, and then cut medallions from the rolled lambchetta for his butcher case.

Boning out the loin, with the flank still attached.

Removing the tenderloin.

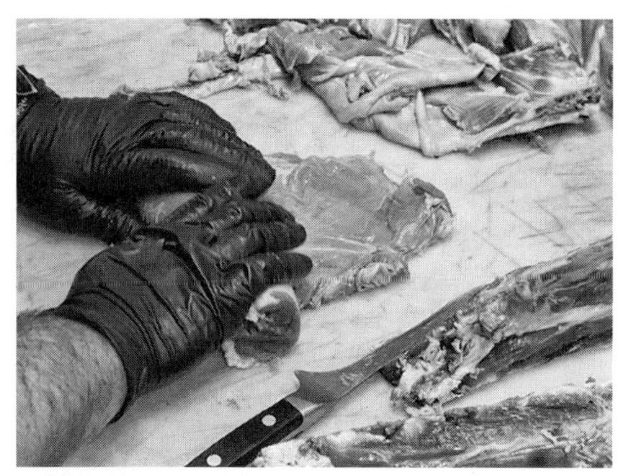

Rolling up the lambchetta.

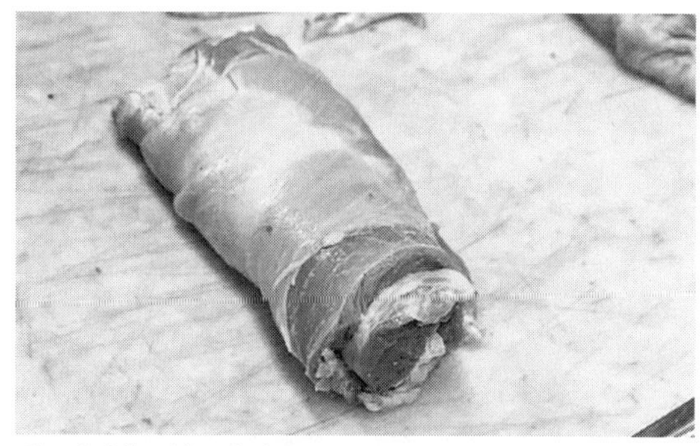

The finished lambchetta.

If you cut medallions, I'd recommend marinating the tied cuts overnight, and then grilling them.

For the other half of the saddle, we'll be cutting bone-in porterhouse portions. You've often heard porterhouse referring to beef; beef porterhouse steaks include the strip loin (New York strip) and the tenderloin. The difference between T-bone and Porterhouse steaks relates to the amount of tenderloin left on the steak. At the thinner end of the tenderloin, you get T-bones. The thicker head of the tenderloin left on strip loin steaks produces porterhouses. With lamb, it's the same. To begin, Matt removes the flank entirely and adds the meat to the grind pile for sausage.

Trim up the bone-in loin, and remove that 13th rib from the end.

Mark porterhouse steaks about two fingers wide, using your cimeter to cut until you reach the bone.

Finish knocking off each chop with the cleaver and mallet.

Removing the flank meat.

Remove the 13th rib, and any trim from the loin end.

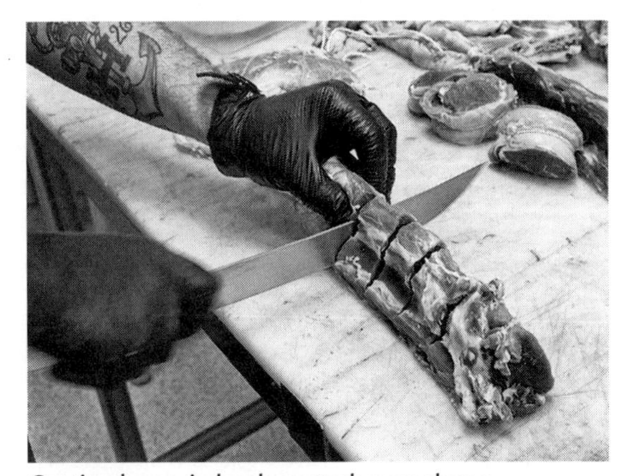
Cutting bone-in lamb porterhouse chops.

The Leg

Finally, it's time to deal with the leg primals. Again, you'll start by splitting down the center of the spine.

Your best options with leg meat will be whole bone-in leg of lamb, boneless leg of lamb, bone-in or boneless leg steaks and sausage meat. We will leave one leg bone in and bone out the other, to prepare you for all of these options.

First, remove the hind shank, just above

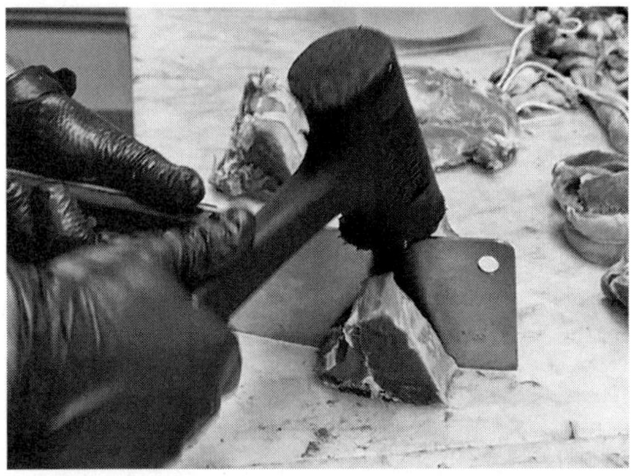
Using the mallet and cleaver to finish cutting porterhouse chops.

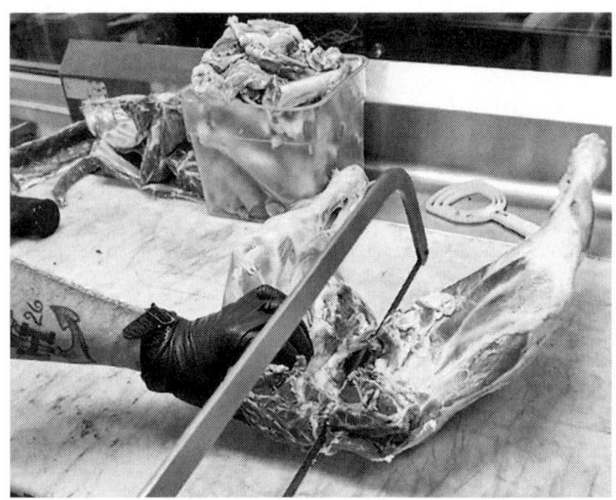
Splitting the leg primals.

The finished bone-in porterhouse portions. You can see that the steak on the far right has less tenderloin. This is moving into T-bone territory, and the rest of the steaks we produced from this strip loin would properly be called Lamb Porterhouse steaks.

The two legs, separated.

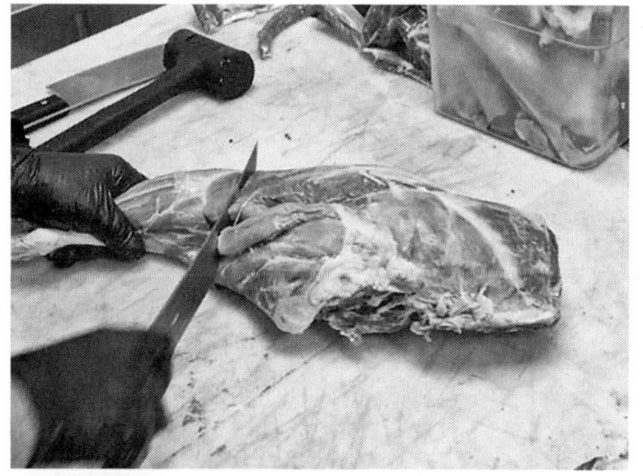
Begin with the cimeter to remove the hindshank.

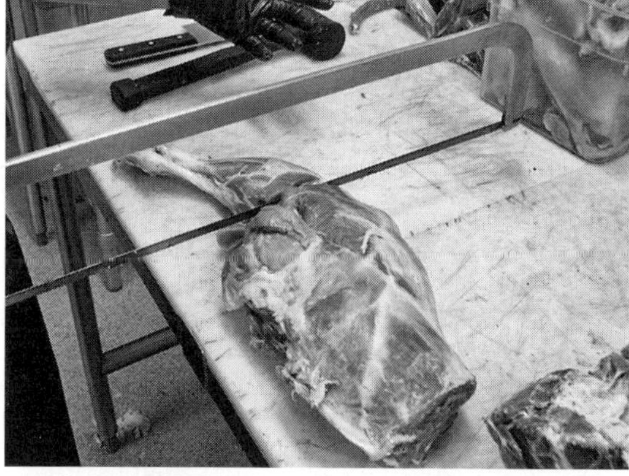
Finish removing the hindshank with the bone saw.

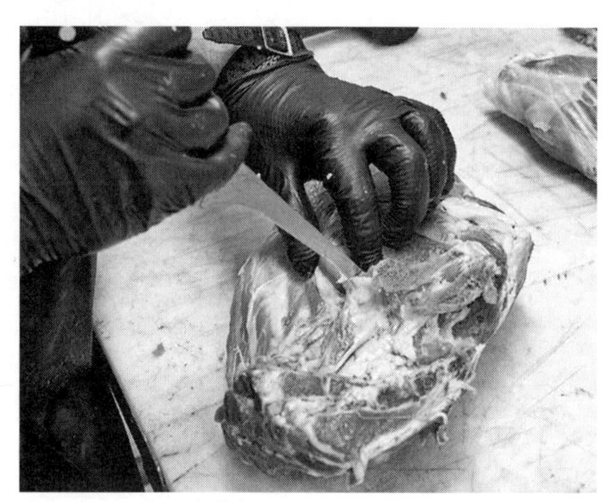
Cut under the aitch bone.

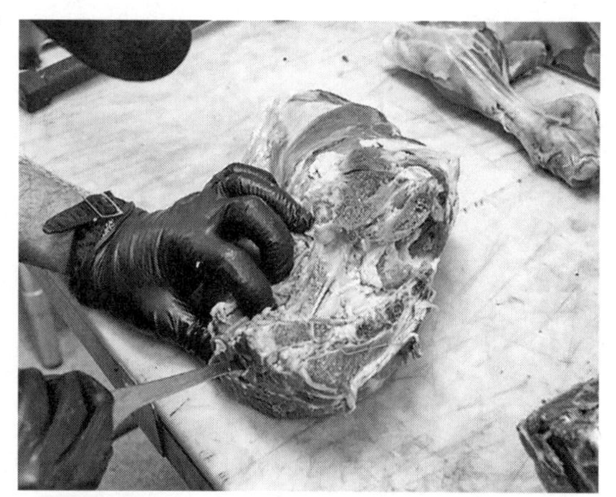
Removing the aitch bone.

the knee. Matt begins with the cimeter, cutting around the bone, and finishes with the bone saw.

To begin boning out the leg, cut under the aitch bone (hip bone), and carefully separate it from the muscle. You'll have to rock your knife around in the socket joint, where the femur joins with the hip. You'll use the joint movement to help you separate the bones.

Now you should be able to see the end of the femur bone from where you removed

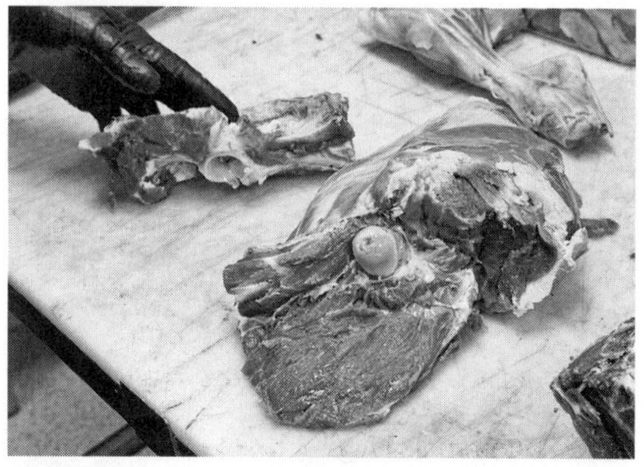
The exposed femur (the freed aitch bone is to the left, in Matt's hand).

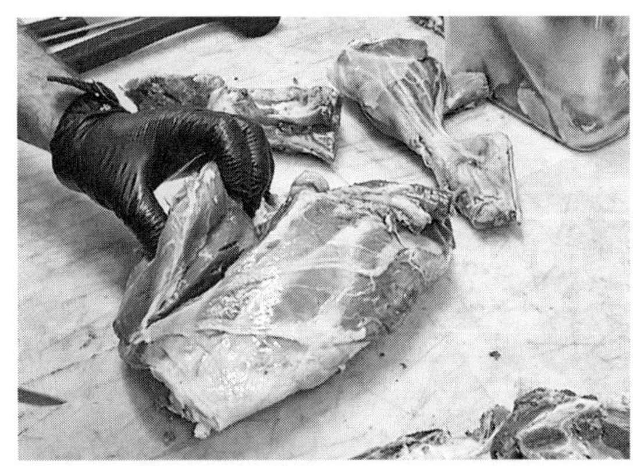

Following the line of the bone, cut straight down into the leg to expose the top of the femur.

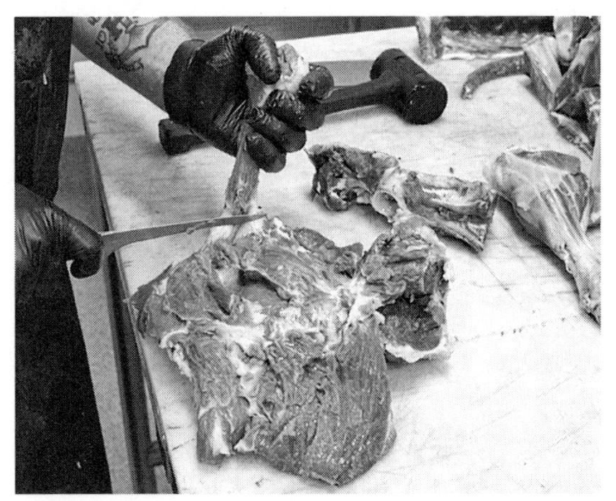

Removing the femur bone.

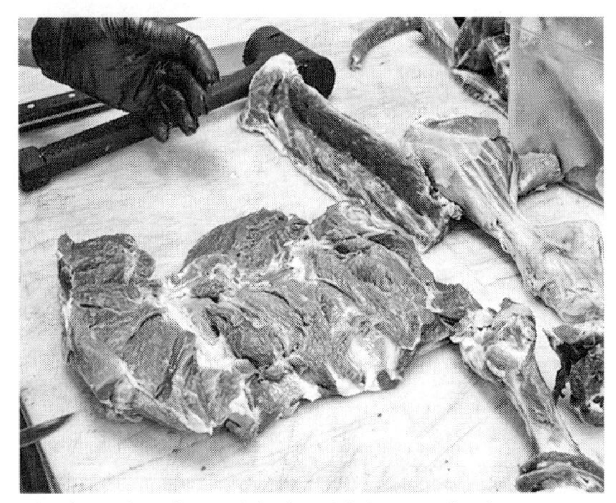

The boneless leg, with the sirloin attached.

the aitchbone. Make a straight cut from the middle of the leg along the top of the femur, exposing the bone. Then cut around the bone, leaving as little meat on the bone as you can.

Before you now is the boneless leg. The sirloin is attached still. In the photo below left, you can see it on the top left. If you need extra lean for your sausage grind, you can take the sirloin off. You can also remove it, butterfly it, and grill or fry it for a delicious serving. Curing and smoking it would be another option. Here, we will just fold it into the leg, and roll it up inside the boneless leg roast.

Once you've seasoned the leg and folded the sirloin in, you'll have a pretty even piece for rolling. Roll up and place twine every two or three inches.

The other leg, which we left bone in, can be smoked whole, with the hind shank still on, or you can remove the hind shank as you did with the boneless leg. If you want to cut bone-in leg steaks, remove the shank and the aitchbone, and then mark steaks with your cimeter. Use the bone saw to get through the femur. You'll end up with large, round steaks with the cross-

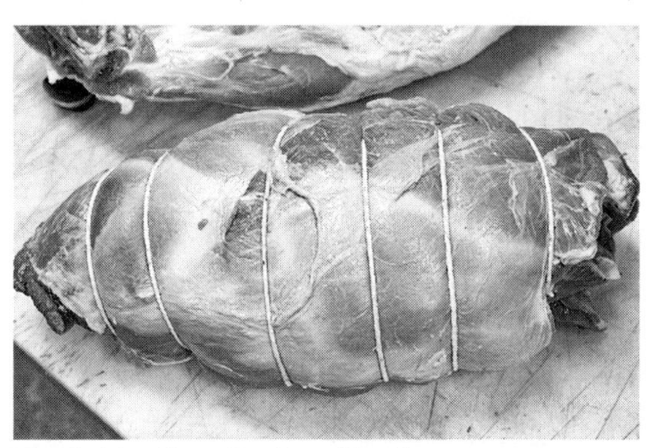

The finished boneless leg of lamb.

cut femur bone in the center. Leg steaks are suitable for braising on the bone. If you bone out the leg, you can separate its muscles and butterfly them for frying or grilling, marinated.

Cooking With Lamb

Lamb is a complex journey in taste, which makes it an adventurous chef's playground. It has wild, earthy notes, sweet undertones, and extremely flavorful fat. Depending on how you want to present your dish, lamb gives the cook a canvas for tart, peppery, fruity, sweet and herby creations. Sometimes, when I create menus, I imagine I am charged with growing a garden around a specific theme, and the only things allowed in the garden are ones that will best showcase the theme. When lamb is the theme, the garden in my mind is colorful, fragrant and varied.

Lamb being a young animal, you will want to take care not to overcook. Many people have had less than pleasant experiences with lamb, but upon questioning will admit they felt it was improperly prepared. Let your meat thermometer, and your senses, guide you.

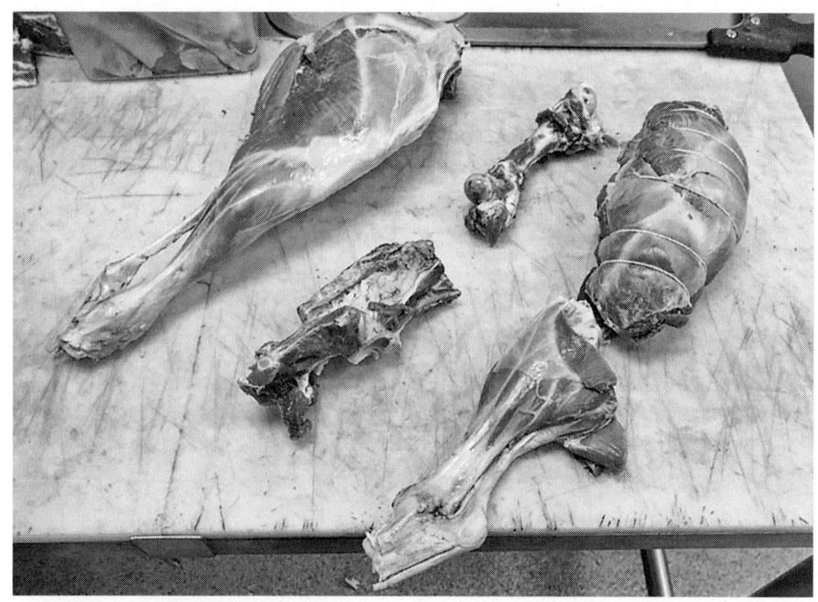

Our results from the leg primals. Clockwise, from top: the whole, bone-in leg with shank attached; the femur; the boneless, tied leg roast; the hind shank; and the aitchbone.

Earl Grey Braised Lamb Shank with Herb Dumplings

This is a lightly hearty meal that combines bergamot and saffron to bring out the sweet undertones of lamb meat.

FOR BRAISE

4 lamb shanks, bone in
1 large sweet onion, chopped
2 celery ribs, chopped
3 cloves garlic, peeled and sliced in half lengthwise
1 bay leaf
3 cups brewed earl gray tea
1 cup dry white wine
Salt and pepper
½ stick unsalted butter

FOR DUMPLINGS

1 cup all purpose flour
½ tsp salt
2 tsp baking powder
1 Tbsp lard
½ cup well shaken buttermilk
4 Tbsp chopped fresh mixed herbs such fennel, sage and oregano
Generous pinch of saffron threads, toasted slightly in a dry pan, until aromatic
Chopped, fresh parsley
½ cup heavy cream (optional)

Preheat the oven to 250°F. On the stovetop, in a large cast-iron skillet, melt the butter over medium-high heat. Add onion, celery and garlic. When vegetables are soft, add the tea and white wine. Bring to a boil, then simmer. As the liquid simmers, brown the shanks. Pat them dry and sprinkle with salt and pepper. Place them in a hot skillet with a small amount of grease, and brown on all sides. Add the browned shanks to the simmering braise liquid, seal with a tight-fitting lid or foil, and transfer to the oven. Braise until the meat is very tender and falling away from the bones.

While the shanks braise, make the dumplings. Toast the saffron. Combine the flour, baking powder, salt and herbs in a small bowl. Cut in the cold lard. Refrigerate this mixture until you are done braising the meat.

Remove the braised shanks from the oven and set the bones and meat aside in some foil or a closed ceramic container to keep warm. Strain the braising liquid into a saucepan. Place the saucepan on the stovetop and bring to a simmer. Taste it. Add salt and pepper if you like. Add the parsley. If you want it more concentrated, cook it down by simmering it for a time. If you're using cream, add it after you've cooked the sauce down.

Now pull your dumpling mix out of the refrigerator and add the buttermilk. Combine until you have a soft dough. Once the liquid in the saucepan is lightly simmering, pinch off one-teaspoon-sized balls of the dough and gently place them into the liquid. Once you've added all the dumplings, put a lid on the pot, turn the heat down until the liquid is barely simmering and allow the dumplings to steam, about 10 minutes.

Serve the lamb shanks with the dumplings and sauce overtop.

Lime-Cream Curry Lamb Sausage with Dosas and Raita

32 oz./2 lb. lamb lean trim
24 oz./1½ lb. lamb fat trim
1 oz. kosher salt
0.2 oz. black pepper
0.5 oz. whole fennel seed
0.5 oz. whole coriander seed
0.2 oz. whole cumin seed
0.2 oz. ground turmeric
0.2 oz. whole fenugreek
0.2 oz. granulated garlic
0.07 oz. ground cayenne
¼ cup cold heavy cream
¼ cup fresh-squeezed lime juice

In a small coffee grinder or spice grinder, combine the fennel, peppercorn, coriander, cumin and fenugreek, and process until ground. Combine with the turmeric, cayenne, salt and garlic. Mix well.

In a small bowl, combine the lime juice and cold cream, and whisk. Slowly add the curry spices, whisking as you go to prevent clumping. Once the curry mixture is well combined in the lime cream, you're ready to season the meat.

In a large bowl, combine the lamb lean and fat trim. Add the curry lime cream and massage into the meat until well coated. Next, open-freeze the seasoned trim and your grinder parts until you're ready to make the sausage.

Grind with the coarse plate, and then run half of the product back through for finer processing. Stuff sausages per the directions on page 166.

These sausages are fantastic served with dosas, raita and fresh spinach and mint.

Dosas

Dosas are Indian crepes made from fermented batter. They are unique, beautiful and flavorful, and a great addition to a diverse diet. Traditionally they are stuffed with a filling and then folded or rolled, but here I am advocating their use as a clever bun for your curried lamb sausage.

1½ cups white basmati rice
¾ cup black lentils, or half mung beans/half red lentils
2⅓ cups water, divided
Salt
Butter or neutral vegetable oil for cooking the dosas

In a bowl, soak the rice in water overnight. In a separate bowl, soak the black lentils (or mung beans and lentils) in water overnight.

The next day, drain the rice and then puree it in a food processor with about ½ cup of water for about 5 minutes. Transfer to a bowl. Next, drain the black lentils and puree them with about ¾ cup of water for 5 minutes. Add to the bowl with the rice. Add ½ tsp salt. Cover the bowl and allow the batter to sit in a warm place for at least 2 hours. You will notice it expanding as it ferments over this time period.

The next morning, stir the remaining water and ½ tsp additional salt into the batter, and allow to rest for an hour or two.

To cook the dosas, heat a generous amount of the butter or oil in a cast-iron skillet until sizzling. Add about ⅓ cup of dosa batter, then quickly thin and even the batter with the back of a spoon. Cook about 4 minutes (if you try to flip it sooner it will stick to the pan) or until you can easily get a metal spatula under the dosa to flip it over. Flip it and cook the other side.

Keep the dosas in a warm oven until you're ready to serve.

Raita

Raita is yogurt sauce. It can be made a million ways. The base for this recipe is simple, since the sausage has many spices involved, but you can vary as you wish.

1 cup full fat, plain yogurt
¼ cup chopped, fresh cilantro
¼ cup minced onion
Salt to taste
Optional additions: cucumber, coriander, lime juice, ginger, etc.

Mix all ingredients in a bowl and season to taste with salt. Chill until ready to serve.

Fire-Cooked Lambchetta with Apricot and Rosemary

A true showstopper. Use fresh rosemary for best results.

2–3 lb. lambchetta (see butchering instructions for details)
½ lb. unsulphured, dried apricots
½ cup sherry vinegar
1 cup water
5–6 cloves garlic, minced
5 Tbsp fresh rosemary, ground in a coffee grinder or spice mill
Zest and juice from one large lemon
2 tsp red pepper flake
Plenty of kosher salt for salting the wrap
Butcher's twine
Cotton cloth, or fine-woven muslin or cheesecloth

Unroll the lambchetta and pat it dry. Score the skin of the belly, and make a few score marks in the loin. Sprinkle liberally with salt and pepper, and allow to rest while you prepare the apricot filling.

Place the apricots and water in a saucepan and simmer over medium-high heat, allowing the fruits to soften. When there is liquid only barely surrounding the apricots, add the vinegar, remove from heat, and place the mixture in a small food processor. Puree. Transfer the pureed mixture into a bowl and add the remaining ingredients.

Now rub the pureed mixture liberally on the inside of the lambchetta and roll it up slowly as you go, beginning at the loin. No need to rub marinade on the outside of the roast, unless you want to. Place the roast on a plate, cover and refrigerate overnight.

When it's time to cook your roast, let it sit out and come to room temperature. Get your fire going; gather up your friends; grab some good wine. Roll out your cotton cloth, doubled over. Salt it generously with kosher salt. Place the lambchetta at the end of the cloth and roll it up, so the meat grabs the salt as it is wrapped. Tie the package with butcher's twine, and place it right in the fire's coals. That's right. In about 20 minutes, start poking your meat thermometer right into the center of the lambchetta. Ideally, you'll want it around 125°F or 130°F when you pull it off the coals. Unwrap it (you may have to cut off the cloth and salty crust) and let it sit a minute before slicing thinly and serving.

Roasted Lamb Rib with Orange, Fennel and Honey Marmalade

Use the full rib for this, with plate meat still on, or you can use this recipe for a frenched rack.

MARMALADE

- 4 oranges, sliced thin with peels still on
- 1 whole fennel bulb and ½ of its leaves thinly sliced (you may choose to use fewer leaves)
- 1 lemon, zested and juiced (reserve both) and then quartered with its seeds
- 6 cups water
- 1½ cups sugar
- 1½ cups honey

Add the oranges, lemon zest, lemon juice and water to a pot. Bring to a boil. Meanwhile, wrap the quartered lemon with its seeds in cheesecloth and tie to make a package. This will provide pectin to help the marmalade gel. Add the lemon package to the boiling pot and reduce to a simmer. Simmer up to an hour, until the oranges are very soft.

Place a small plate into the freezer. (You'll use this to test the marmalade's consistency.)

Once your fruit is sufficiently softened, reboil the mixture and add the fennel, sugar and honey, then stir constantly until the mixture reaches 220°F. If it's a tad under, don't worry. This step will take about 30 minutes.

Remove from heat. Take the plate from the freezer and place a spoonful of marmalade on it. After a moment, tilt the plate. If the marmalade wobbles and slowly slides, unified, you're done. If it is runny and watery underneath and clumpy in the center, you'll need to stir and heat until it is further gelled.

When the marmalade is done, transfer it to jars while hot, and seal. This stuff is great on toast, pastry and biscuits, so make plenty more than you'll need for your lamb rib.

FOR LAMB RIB

- ¼ cup neutral vegetable oil, like sunflower or grapeseed
- 2 cloves garlic, minced
- ¼ cup sweet paprika
- 1 tsp fresh ground black pepper
- 2½ tsp salt
- ½ cup orange, fennel & honey maramalade (above)
- ½ cup red wine vinegar

Combine the oil, garlic, paprika, pepper and 2 tsp of the salt in a small bowl. Pat the lamb rib dry and rub it thoroughly with the spice mixture. Place the rib on a plate, cover and let sit in the refrigerator for at least six hours. Before roasting, let the rib sit at room temperature for about an hour. Preheat the oven to 350°F.

To roast, transfer the rib to a heavy shallow baking dish and cover the dish tightly with foil. Bake the rib for 1¼ hours. As it bakes, stir together the red wine vinegar and remaining salt in a small saucepan. Heat to a simmer. Once your lamb rib has baked, remove it from the oven and uncover it. Pour off the fat from the pan. Brush the rib with the marmalade mixture and return to the oven, uncovered, for 10 minutes. Remove it, flip it over, brush the other side with the marmalade and return it to the oven, uncovered. Continue flipping and brushing the rib every 10 minutes until the marmalade is gone.

Serve on the bone, by cutting in between the ribs to portion.

Roast Leg of Lamb with Mustard, Capers and Marjoram

I adore marjoram, with its fat, fragrant blossoms that scream of soup or roast beast. Harvest the marjoram with its blossoms swollen, and save some sprigs to serve with the finished roast. Use this rub for a bone-in leg roast, as detailed here, or a boneless, tied leg roast. Be sure to season the boneless roast before tying it, and note that cooking times will differ for bone-in and boneless cuts.

1 leg of lamb or mutton, about 8 lb.
½ cup coarse ground mustard
3 cloves of garlic, minced
½ cup capers with brine
6 or 7 sprigs of fresh marjoram, chopped
2 tsp salt
1 tsp fresh ground pepper
olive oil

Pat the bone-in leg dry, and bring to room temperature. Score the leg in a cross-hatch pattern with your boning knife. Preheat the oven to 450°F. In a bowl, combine all ingredients and rub over the leg, thoroughly. Place the rubbed leg in a roasting pan and roast at 450°F for 30 minutes to brown and form a crust. Then reduce the heat to 325°F and add a scant amount of lamb stock to the bottom of the roasting pan. Roast for an hour or more, until the lamb is at least 125°F. Adjust cooking time according to your liking.

Once the leg is done, rest it at least 15 minutes before carving.

Bourbon and Sorghum Glazed Lamb Spare Ribs

Bourbon and sorghum syrup are two of my favorite things in life. Sorghum syrup is juice from crushed sorghum cane, cooked to caramel perfection. If you find yourself in the shameful circumstance of not being able to find sorghum syrup, use molasses. But understand that they are not the same: molasses is a by-product of the sugar industry. If you find yourself in the unfortunate circumstance of not being able to find bourbon... well, use whiskey, I suppose.

3–4 lb. lamb spare ribs, or combined Denver ribs and spare ribs, cut into two-rib pieces.
2 Tbsp chopped garlic
1-inch piece of fresh ginger, peeled and minced
¼ cup bourbon
¼ cup sorghum syrup
2½ cups lamb stock
Salt and pepper

Salt and pepper the ribs, dust with a tinge of paprika and garlic, and wrap in plastic. Refrigerate overnight. Heat a large skillet that will hold all the rib pieces in one layer over high heat. Add the ribs and the lamb stock and cook until the stock has mostly boiled off, about 20 minutes. Reduce the heat and brown the ribs in their own fat and what little stock remains. Turn them as they brown, and add the garlic and ginger toward the end. Meanwhile, combine the bourbon and sorghum syrup. When the ribs are browned and the garlic and ginger are soft and aromatic (do not burn), add 1 cup more of lamb stock. Increase the heat and allow the mixture to boil down. Once the ribs have darkened and the liquid has reduced away, add ½ cup lamb stock. Once the liquid has boiled away again, add the bourbon and sorghum mixture and lower the heat. Let this reduce until the meat is tender and ready to eat.

Serve over buttery grits, with some collard greens.

Sauces and Sundries for Lamb

Broiled Tomatillo "Salsa"

Tart tomatillos make a great pairing for grilled lamb. I use the term salsa *loosely here.*

1 lb. tomatillos, husks removed, and halved
½ lb. sungold cherry tomatoes, halved
2 poblano peppers, quartered
5 or more cloves garlic, peeled and halved lengthwise
2–3 cipollini, chopped, or 1 medium sweet onion, chopped
Ground cumin
Salt
Fresh ground black pepper
Juice from ½ lime

Place all tomatillos, poblanos and tomatoes skin-side up on a broiling pan. Surround them with the cipollini and garlic, and sprinkle all with cumin, salt and pepper. Broil in the oven until the veggies begin to blister and the garlic and cipollini begin to blacken. Remove from oven, squeeze lime juice over top, and serve.

Red Wine Mushrooms

This recipe has been boosting soups, cuts of meat and salads since I entered my first kitchen.

⅓ cup red wine vinegar
⅓ cup olive oil
¼ cup minced sweet onion
1 tsp salt
2 Tbsp parsley
1 tsp mustard
1 Tbsp brown sugar
2 cloves garlic, minced
1 lb. fresh mushrooms, sliced

Stir all ingredients except for mushrooms together in a saucepan and bring to a boil. Reduce heat, add mushrooms and simmer 10–12 minutes.

Ginger Mint Cilantro Chutney

This is a bright, herby accompaniment for fried lamb noisettes, butterflied leg steaks or lamb sirloin, pounded and pan-cooked.

1 cup chopped fresh cilantro
½ cup chopped fresh mint
2 dried chiles, crushed
1 Tbsp fresh ginger, minced
2 garlic cloves, minced
Juice from one lime
Salt to taste

Combine all ingredients in a food processor and blend thoroughly. Add oil or lime juice as needed to adjust consistency to your liking.

Grilled Artichoke Salad with Smoked Paprika Aioli

Vary the aioli in this recipe to create endless permutations. Another amazing trick is to smoke the red onion first. It's easy. When you've got your smoker going for some other purpose, throw in a couple of whole red onions with the skin on.

FOR AIOLI

- 1 whole egg, at room temperature
- 1 egg yolk, at room temperature
- 1 tsp Dijon mustard
- 2 Tbsp lemon juice
- 2 tsp smoked paprika
- ¾ cup neutral vegetable oil, such as sunflower
- Salt, to taste

NOTE: This aioli recipe makes twice what you need for the salad, so you can save the rest in the fridge for a yummy sandwich, later.

FOR THE SALAD

- 2 lb. small artichoke hearts, rinsed
- 1 red onion (smoked if you're feeling awesome), in eighths
- 6 oz. capers, drained of their brine
- 5 cloves garlic, split in half lengthwise
- 3–4 oz. good Parmesan cheese, grated
- Salt and fresh ground pepper to taste

Heat the grill. Skewer the artichoke hearts, onions (unless you already smoked them) and garlic. Char these over direct heat and set aside.

Prepare the aioli. In a food processor, mix all ingredients except oil and process until well blended. Add the oil so slowly that you think you might collapse. No kidding. Drop. By. Drop. Taste and check seasoning. You may want more lemon juice or paprika.

In a serving bowl, combine the charred artichoke, onion, garlic, capers and aioli. Stir to coat all veggies. Top with grated cheese and serve at room temperature.

Pork

If you have spent any time in rural spaces, you have heard the saying, "A person is rich if he has land." I maybe heard it first in school, in some text about Thomas Jefferson, and then heard it again in the deep Alabama drawl of my granddaddy, and then again from the mustachioed mouths of farmers when I moved to the country, to Old Fort, North Carolina. I heard it many, many times, but I never learned it like I did from pigs.

I grew up in the city, in a small apartment with my mother and brother, and I did not touch much earth, early on. I think the earliest and closest I ever got to the ground was in jumping from a tall tree house, in elementary school, so high that you could not come down without biting your tongue. Later, my mother did a lot of camping with us, and it was then that I first heard the giant hush inside of me. I didn't know then what it was.

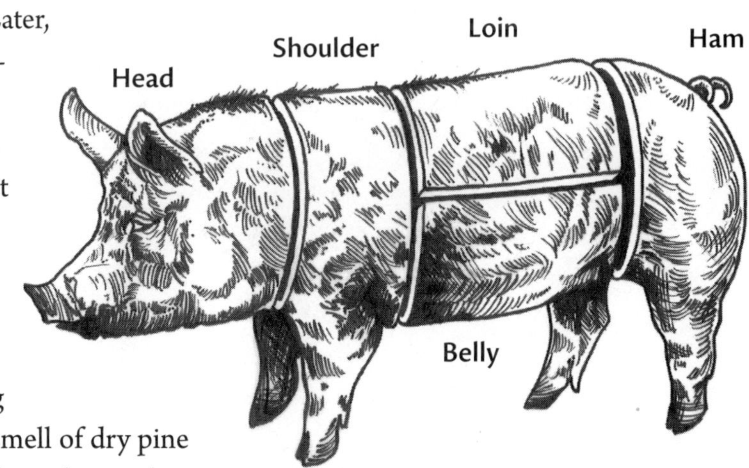

Now, I'm sitting under a tree, among a scattered group of foraging heritage pigs. The sun is filtering through the leaves, and the smell of dry pine and wet earth is even in my hair. I'm twirling a pinecone between my fingers; I'm getting dirty; and I'm squinting through the trees. I'm thinking about the land, watching the pigs as they love it, thoroughly, snuggling themselves into every inch of the soil's folds, making masks of the cold mats of oak leaves, letting their bellies graze the sweet ground. Owning the place.

I owned land, once.

Looking back, I like to think I can remember every single morning on my farm, but really what I remember is the way that it feels each time to walk out the door into it. That's the hush. Even in the spring, when the birds are loud and ecstatic and the grass in the pasture is a shade of green damn near embarrassing. Even when your head is jangling with four thousand chores and not enough hours, and there is something on the list that you've never done before, that you're not sure you can do. But you walk out through the dawn grass anyway, and you are instantly soaked with it.

I used to feel like an invalid every spring, or like someone who had been asleep, because I was always so surprised at how new the earth made itself. It did not matter how many years I farmed, the hope of another season was infectious and unquenchable. And I was suddenly capable, braver and better again, too. It goes like this when you own land. When it finally rains, you realize how thirsty you've been. When it finally frosts in the fall, you give in, exhausted. And when spring finally comes, you jump up. You're ready. It's emotional. No, it's soulful, to be a person of the land. In many ways, we people are out of practice. We struggle at being so deeply, and so reflexively... earthy.

But the pigs, now, are lying in the sun. It is the stillest, sweetest moment, there in the mud, and they are all spooned up against each other sideways. Every now and again an ear twitches, or a chubby head bobs, but the pigs and the sun are mostly etched there, silently. There is one fellow not sleeping, lolling about on slow legs at the edge of the pasture. Watching him, I recall times when I loathed the land's hold on me. When we stuffed pillows into the skylights before bed to save our ears from the sound of weather. When I felt thoroughly wasted by the day, like I had utterly failed. When the best-laid plans were dashed by some higher design, or my own stupidity. There are some sorry days, farming.

Those days get folded into the body, too. Like pigs, we humans take it all in. How much it shapes us, we may not know. The pig is quickly assimilated. The pig cannot resist the land, and when the pig dies, we eat its body. If we're really eating, we muse on whether the body is enough homage to the land. Whether we can taste the fog, and the seeds, and the fruit. For the better it tastes, and the better it feels, the better we know it lived.

When I die, I hope the land lies in me. I hope I have spent enough days as a pig would. Because I know that when I sold my farm, I mourned. I asked everyone to be gone when I packed up my tools and my plants, and I walked up and down the gravel road, just weeping. I made a few trips back to grab things, but mostly I just left with what I could. It was suddenly a terror to be there, but not in it, and without it. Untied from it. I felt a thousand hushed mornings at once, felt all the failures and all the victories at once, and realized I was wholly lost, no longer anticipating or responding to the land's rhythms. I looked around at many patches of ground I had attempted to cultivate over the years, remembering small wonders and feeling again the smart of my mistakes. But mostly I remembered how hard I had tried. Lord, how I tried. No matter how stupid I had felt at each agricultural loss, or how messy some of my intentions looked on the ground, I had been so hopeful. I had deeply loved.

I realize now how fortunate one is to be a person of the land, the land that forgives. That's just the thing. It gives, and gives, and gives, without loving you back. Its marvel is of its own necessity, and it drives on, with or without your grace. To be truly rich is to feel beauty and wildness every day, and to have the sense that you have a chance, within a larger scheme. To be rich is to be humbled and hushed, over and over. To work, and to become better. To be alive to the world.

And so the last time I visited the farm, I began to walk through all the fields, hearing my footsteps against the emptiness. And suddenly, I felt as though I could not well enough leave the land. I wanted to roll, weeping and smiling in the mud. I, too, wanted to root there. I wanted to eat thousands of pink-white cherry blossoms and buckets of toasty acorns. I wanted to sleep with my face in pine needles, and drink dew, and wake up with my nose frozen, and fall asleep again when the air was so thick with heat that I could barely breathe. I wanted fireflies and snowy feet. I wanted to work, cursing in the rain.

Yes. The pig is rich of the land. She knows how to respond to it, and how to move it, too, not banking on next time. Not taking for granted any tiny bounty, any fragrant gift.

Raising Pigs

Pork is a fascinating farm enterprise. It is rewarding in that pigs are such amicable animals, and the turnaround on finished product is relatively fast compared to ruminant animals like cows and sheep. To the credit of the eating public (and the rendering industry), we make better use of the whole pig carcass than we do any other animal raised on the commodity market. While it can be daunting to work pigs into a multi-species pastured operation, or even a holding that boasts mostly vegetables, it can be done. And when it is done well, the end result is one of the best things on earth. Actually, when I sat down to begin tapping out this chapter, I realized after some time that I had just been sitting and staring at a blank screen, focusing only on the imaginary umami on my tongue, as my entire life in pork (field, kitchen and table) flashed before my eyes. If you seek to raise only one species of animal this year, let it be pig. Trust me.

The vast majority of pork raised in the United States is grown in confinement, on close to 70,000 farms throughout all 50 states. These operations turn out 113 million animals annually, which equates to 23 billion pounds of pork. Approximately a quarter of US pork is exported, while the rest goes to our plates, making it our third favorite meat (after chicken and beef). It's a healthy market, in other words, and small landowners looking to sell or simply eat pork raised on pasture or in wooded lots, from heritage breed animals, can create a product vastly different from the commodity pork on our supermarket shelves.

Interestingly, although the average person could, without too much convincing, more easily eat the entire pig carcass than any other livestock animal we encounter daily, only about 50 percent of the carcass is used for fresh meat in commodity production. From the leftovers, pituitary glands are carefully removed for insulin production, organs are used in the medical field, fat is rendered, and countless other bits and pieces are used in paints, varnishes, makeup, dynamite, toothpaste, leather and more. While there is very little waste in the processing of swine, the manure waste that the growing process produces in our country is crippling, combining with the rest of our feedlot stock to make livestock the largest producer of atmospheric methane.

Piggybacking (huh!) on my aforementioned holistic foodthink, it's worth noting that landfills are the country's third largest producers of methane, and include, by weight, 20 percent food waste. Add to that the dreadful waste incurred in the production of vegetables, cereals, fruits, oilseeds, roots and tubers. Almost half of all fruits and vegetables produced annually are lost or wasted, from seed to home trashcan. This includes plants never harvested, losses in distribution and processing, and finally, food that is wasted in homes. I find this fascinating, especially considering the relative efficiency of the porcine industry—and the fact that most breeds of the wonderful old pig will devour just about any fruit, vegetable, nut, seed, root or tuber placed at the end of their snouts.

I estimate (based on a feed conversion of 3.5 lb. feed to one pound of pork) that the US pork industry feeds a little more than one billion pounds of feed to pigs annually. Let's imagine, based on Food and Agriculture Organization food loss research on cereal crops, that 30 percent of that feed is lost *before* it is delivered to the farm gate, either due to weather events, distribution issues, processing snafus, or whathave-you. Does this mean we are producing and wasting, for purposes of bland-pork production, upwards of 139.5 billion pounds of cereal and cereal by-product, all while 133 billion pounds of vegetable, fruit and cereal trash is yearly thrown into landfills? Fascinating. Fascinating indeed. It is obviously no small measure to take this disturbing mathematical equation and put it into action. We cannot take some big, well-meaning scoop, gather up all that waste and deliver it to pork farmers in 50 states. I know a little bit more about farming and industry than that. But wow.

I also happen to know that of vegetative foods consumed in the United States, about 7 percent are usable or converted by the body once digested. Yes, I am talking about the human feed conversion ratio, your body's efficiency at converting calories to energy, just as a pig does—and doing so more easily with meat, nuts and avocadoes than with celery and tomato. What a wonderful illustration of the holistic, systems thinking that we can be doing about our food choices. And what a fantastic argument for your own jolly herd of home pigs. Turn off your garbage disposal.

This section will concern the raising of pastured pork, and how to incorporate pigs into your diversified homestead. Making this effort will bring you joy, delicious food and powerful land management capabilities.

Breeds

Typical American pork comes from pigs bred to reproduce in quantity, grow fast and stay lean. Because of this breeding, and the cocktail of corn, soybean, antibiotics and additives we feed to our swine, we've sacrificed the product for the market, so we're turning out plenty of piddling, weak muscle and a compromised eating experience. While many larger pork farmers have now turned to heritage breeds such as Berkshire and Tamworth, the move to fattier, more flavorful breeds is only half the battle. Pigs are omnivores that enjoy as diverse a diet as we humans do. They also need and want to move. They are strong, happy animals. Deny them the ability to root and tumble, feed them a weak diet, and you will produce meat for sure—but not strong, happy meat.

Without a doubt, raise heritage breeds. The best pork I have worked with comes from old fatty breeds such as Guinea crossed with good-breeding, muscled varieties like Hereford. Research the history of breeds while remaining mindful of your intended use, and your climate. For example, pigs like the Ossabaw (one of my favorites), which developed in island climates with less bulky forage, will have higher fat stores and peak resistance to warm-weather diseases and parasites. They may grow more slowly than a more muscular breed like the Duroc, so a cross can produce an animal with a great balance of high-quality fat and superior lean. If you plan to farrow (keep breeding sows and raise babies), also include a breed with good mothering characteristics. On my farm, we raised quite the mutt heritage pig, with lineage from Tamworth, Berkshire, Hampshire and Old Spot throughout our herd.

With respect to parasites, disease and climate, selection of pork breed is nowhere near as delicate as in lamb production. And when it comes to the animal's performance on pasture or open range, breed consideration is nowhere near as confounding as beef production. So, by all means, strive for superior flavor when building your pig enterprise. Farrow-to-finish operations (wherein pigs are born and fed to slaughter weight on the same farm) can be difficult for beginners, so the best place to start is by purchasing some weaned youngsters, called *feeder pigs*. That being said, there is a real demand in the homestead and small farm community for well-raised feeder pigs, so if a farrowing operation is of interest to you, explore the possibility of an enterprise that offers strapping young heritage feeders to interested farmers and landowners.

Genetics, no matter the animal, tend to fascinate me. But while I love to play with ways of combining the characteristics of several heritage breeds, I am also well aware of the importance of genetic integrity. When it comes to swine, small landowners can play a big part in preserving genetic diversity by focusing on working with purebred lineages. If you are considering farrowing, look at purebred lines and consider registering your stock with the Livestock Conservancy.

Space and Water

There are many ways to raise pigs outdoors, and the best course of action will relate to your climate, soil type, existing vegetation and overall farm goals. Pigs are omnivores and active, social animals, so they will do best with space, movement and variety in their environment. I'm most excited about *silvopasture* (livestock raised intentionally in woodland areas) and intensive rotations of swine with both annual and perennial food or forage crops. A combination of all of these approaches could be used on a land base that boasts both open pastureland and forest, or forest edge.

No matter your plan, the challenge with pigs is to balance stocking density and time on the ground, to avoid nutrient overload from manure and excessive soil damage. The natural tendency of the pig is to use its strong snout to root through the soil for grubs and seeds, so its impact on the soil is much greater than that of ruminant animals. You've seen pictures of hogs with rings through their noses—the rings are there to prevent rooting. There is no need for this type of cruelty. The animal should be able to exhibit all of its natural tendencies, and the farmer should have a proper rotation plan and a strategy for using those tendencies to her benefit. Some breeds root more lightly than others, so if you are dealing with a small space, choose smaller-bodied breeds to lighten their impact on your land.

Figure a minimum of 1,500 and a maximum of 3,000 square feet per animal. You can certainly stock more intensively, but if you do, you will need to move the animals more often. They will eliminate ground cover quickly and establish wallows, large craters in the ground where they can loll about in the mud and rain. If you are following an intensive pasture management system and attempting to establish perennial forages with minimum soil disturbance, the pig is probably not your ideal species. However, if you have sufficient land to establish large enough paddocks, can stock the animals less densely, and establish a plan to sow perennial grasses in the paddock directly after the pigs are removed, you can then give that area a break from animals and grow green cover crops eight months every year and achieve your goal. This type of system would potentially be ideal for a grower who eventually wanted to re-seed, or very intentionally direct the forage mix in his or her pasture.

Swine can be raised more intensively in rotation with annual crops, giving paddocks two to five months of rest in between stocking. This is the method we used on our farm. Pigs were fenced into an area, allowed to feed, wallow, root, and snuggle, and then just after rotating, we would sow a fast-growing cover crop such as sorghum-sudangrass, buckwheat, millet, oats or rye (depending on the season). This worked best in the summer months. We would grow the cover crop to maturity and allow the cattle to graze it down. By then the pigs would have moved onto other areas, so the process would begin again. We did not perfect this system by any means, but I think if we could have rationalized total land management priorities between all our row crops and animal species, and managed our plan from season to season, we could have produced an incredibly dynamic rotation.

Consider, for example, the idea of rotating 25 or so pigs in a little more than one-acre paddocks. This gives you roughly 2,000 square feet per animal in each paddock. Once they've reduced ground cover by 30 to 50 percent, move them to a new paddock, with a slight vegetative buffer between the new paddock and the old. Immediately sow a season-appropriate cover crop. In spring, perhaps you'd plant oats, increasing your seeding rate to ensure quick and reliable ground cover. As the season progresses, vary your crop to correspond with the design that follows. Perhaps in the next paddock or group of paddocks, you'll follow the pigs with buckwheat (early summer); in the third, sorghum-sudangrass; and in the fourth, millet or barley. You could flail mow each cover crop at maturity to kill it, leaving the residue to build soil organic matter. The last or second-to-last paddock that the pork visits before slaughter could be sown with a cover crop that will winter kill (meaning the first frost will wipe it out). Leave the mulch and dead stems until spring, then plant through them with a garden crop such as potato or cabbage in the spring.

Other ideas include cutting the matured cover crop for hay, or using it as a nurse crop for a brassica or forb that you can use to graze your sheep. Perhaps many readers will be dealing with much smaller land holdings. This is fine. You can easily take the notes above and figure out how to subdivide paddocks with poly-wire fencing, thereby allowing for smaller paddocks and fewer animals.

The consequence of improper paddock management and stocking when rearing pigs is nutrient buildup, specifically of phosphorus and nitrogen, and subsequent leaching of those nutrients into waterways. This is not a good thing. Erosion is also a concern if pigs are allowed to decimate plant roots that secure the topsoil. Pay attention to your slope, aspect and climate. Do you have enough rain to establish a quick cover after moving animals? If not, provide overhead irrigation once you've sown the crop seed. If you are in a climate where groundcover is already sparse and difficult to establish, perhaps you should consider paddocks in the woods, or reconsider those pig farming dreams all together. Lastly, pigs left to totally upend a piece of ground, and cover it with manure, will create an overgrowth within bacterial populations in the soil, which will support the establishment of weeds. Working with the land is the ultimate goal, while establishing animal systems that complement and enrich the rest of your setup.

Including a wooded paddock in the rotation is an excellent alternative. The famous Iberico ham is raised in Spain in silvopastoral systems called *dehasas*. I can say from experience that it produces superior product in temperate climates as well, allowing swine to dine on mast and tree fruit. You can time the rotation to allow access during acorn season or just after the persimmons drop. This type of management, allowing for a diverse and natural diet, is what gives pork its unique flavor. This is true *terroir.*

Watering systems should give animals access to clean water at any time. Pig drinkers with nozzles activated by the animal's snout are available from several livestock equipment providers, noted in the Resources section. Pigs also dearly love a spray with the garden hose on a hot day, so quick-attach hoses are a great asset in the pig lot. Locate water centrally, so that if you end up subdividing paddocks, you can do so without having to set up additional troughs. And as always, make sure there is a vegetative buffer between paddocks and nearby streams.

Fencing

The most effective fencing for pigs I've used is a three-strand electrified poly-wire fence, secured at the corner with metal T-posts and secured with fiberglass in-line posts. Some people use two wires, but I have found that young feeder pigs can slip through this setup. It would be possible to start with three and remove the lowest wire after the animals fatten up a bit. Poly-wire is preferable to electrified netting for pigs, as they tend to root the soil up against the netting and ground out the electric signal, or knock the fence down entirely. I've spent many cold mornings kicking upturned soil off of the bottom of fence netting with my boot. The poly-wire strands are much more effective, since they are up off the ground at least six inches.

Use solar- or battery-powered fence chargers for your pig fence, or plan to tie the paddocks into a high-tensile perimeter fence. If your paddocks tie into a larger fencing setup, be aware that any interruption of signal in the pig paddocks (due to rooted soil or vegetation touching the lower wires or netting) will drain on the entire system, and you could have a mass, multi-species exodus on your hands if you're not vigilant.

Feed and Minerals

The best pork comes from animals with colorful diets. While many people feed pigs almost exclusively with vegetable and garden scraps, I'd urge you to supplement. Protein is important in muscle development, especially if your pigs are not moving around as much as would be ideal. Organic swine pellets are a good investment, and if you are able to provide additional protein sources like acorns or whey, the animals will consume less purchased feed. Pigs fed slop alone will be flabbier, with weaker flavor and less muscle.

Feeding your hogs exclusively on grain can get quite expensive, so look for additional feed sources, and by all means feed scraps. A typical pig can consume close to a thousand pounds of feed before reaching slaughter weight, so you will benefit from creative rotation and attention to seasonal wild forage. On our farm we used to get the kids to collect the acorns and persimmons that fell around the garden and driveway in autumn. Certain breeds will not fare as well on garden scraps once they have had a taste of grain or high-protein fodder, so it may be worth timing your introduction of different feedstuffs so that your pigs don't get snobby. They are smart and full of personality. They really will test you.

Using organic grain supplements avoids genetically modified feeds, but organic feed is more expensive. Many small farmers are now using spent brewers' grain as a feed supplement. Be cautious in rationing and timing, however, as there is much to be understood about the quality of wet grain over time, and its effect on the animal's overall gastronomy. There is also increasing regulatory pressure on the transport and handling of grain, so cooperation between the brewing and farming community is important. More research is needed to demonstrate the viability of the grain as a feed source and to show that small business has the resources to handle it as it ages; otherwise we'll end up with another valuable animal feed source getting dumped in the landfill.

Sprouting raw grains for pig feed is an excellent option for supplementation and provides high-protein, high-fiber, flavorful food. Protein, fiber and other nutrients do increase in the seed during the process of germination, and palatability can also increase, depending on the seed type. You can sprout grains for all your livestock, but pigs are good animals to use as you're getting started; because they will eat other things, you won't have to worry so much about the volume of sprouted feed in relation to their body weight. Common sprouting grains are barley, wheat, oats and corn. I had a nice little operation of sprouted sunflower seeds for a while. See the sidebar for information on Sprouting Grains For Feed.

As with any livestock, provide free choice minerals for your pigs. The animals will regulate their intake as long as you provide a source.

Pork Butchery

The pig carcass is first split in half, and then split into the following primals: head, shoulder, loin, belly and leg. For our pig breakdown,

Sprouting Grains For Feed

Sprouting barley, wheat, corn, oats or sunflowers for animal feed is a simple and fast process. Start with untreated, organic seed in bulk, and soak in buckets overnight. The next day, drain the seed and arrange it 1–2 inches thick in seed or bread flats, or drill holes in baking sheets to provide drainage. Place the flats in indirect light and ensure room temperature or a bit higher. You don't want a greenhouse-type situation, as too much heat will encourage molding of the seed. Place the flats on a rack or some other setup that ensures airflow on all sides and allows for good drainage. Water the seeds daily, and allow the flats to drain. In as little as one week you can have some sprouts! You can feed them to the animals as soon as the seed opens and produces a radical (the plant's first root), or you can wait until a substantial root mat has developed and the first leaves have formed. Feed the animals the entire thing: seed, leaves and roots. Start new trays of seed every two days or so to keep a steady supply of sprouts growing. ◆

I give you the wizardry of Tyler Cook, head butcher at my former shop, Foothills Deli and Butchery in Black Mountain and Asheville, NC.

Tyler begins by removing the leaf fat. This special fat surrounding the animal's organs is unique and suppler than any other fat on the hog. Best to render it separately for perfect pastry lard (see recipe p. 151). Pull all the leaf fat from the belly and saddle. You should end up with a generous armful, and a newly revealed layer of flap muscle on the carcass.

Next, we'll remove the tenderloin, which is nestled up under the spine, cradled by the ribs. Cut close to the bone, rolling the muscle out as you go. You can leave the tenderloin whole or cut medallions. As with all tender muscles, this cut is best roasted or grilled, using high, dry heat.

Next, we'll remove the leg primal. Tyler begins by cutting back part of the saddle, to give him better access and to prevent his nicking the saddle as he cuts through the hip. Following the outline of the leg, cut back a section of the saddle and fold it over onto the belly before isolating the leg primal.

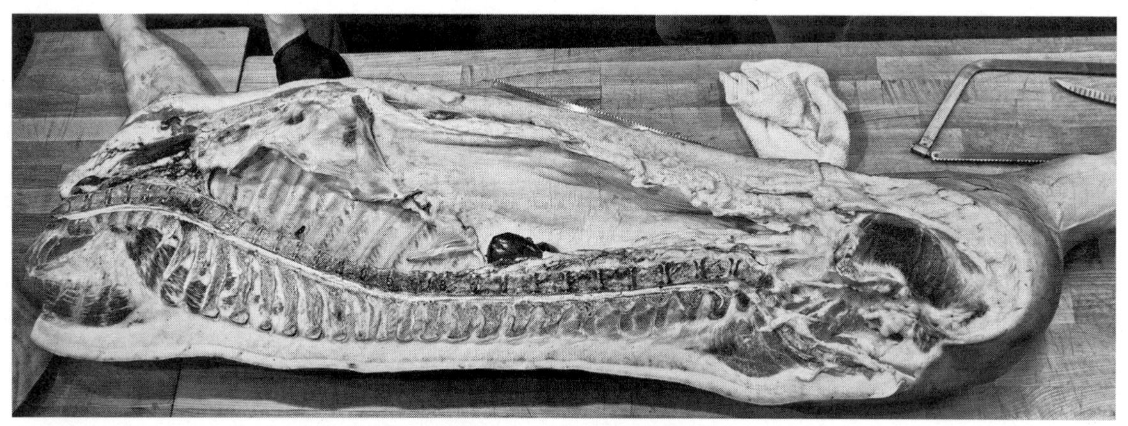

Our half-pig, born and raised at Wild Turkey Farms, China Grove, NC.

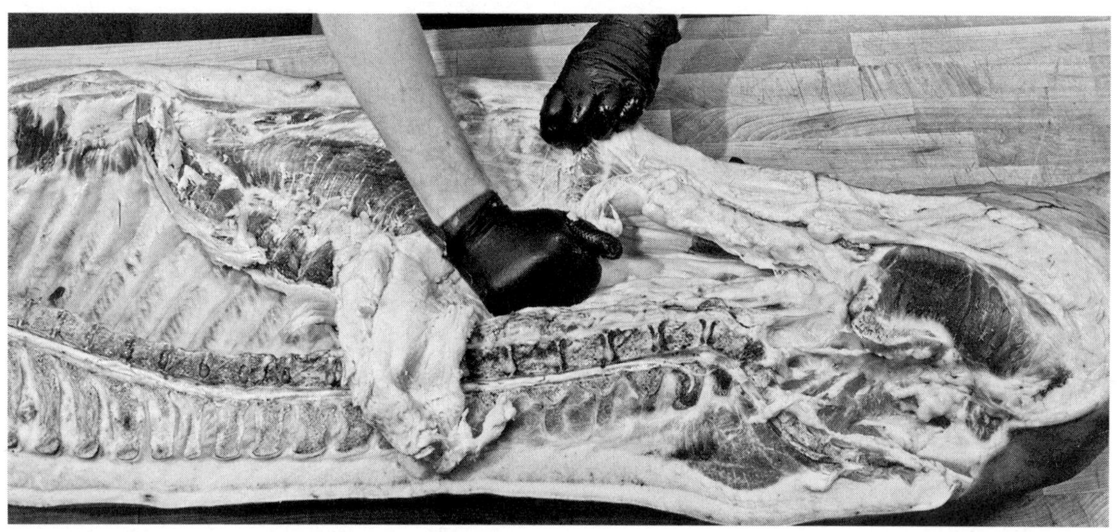

Removing the leaf fat, revealing flap muscle underneath.

Removing the tenderloin.

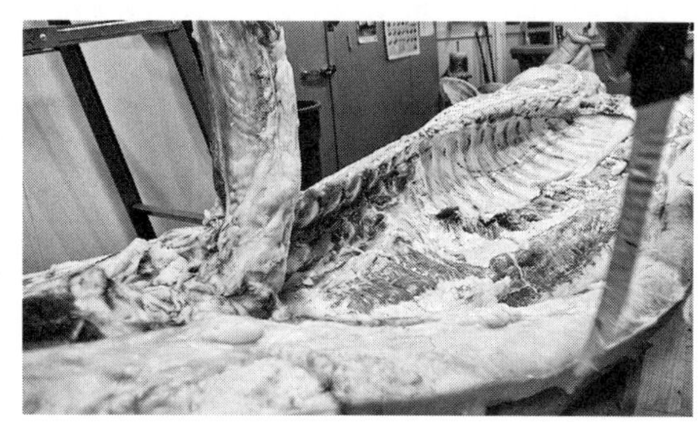

Removing the tenderloin.

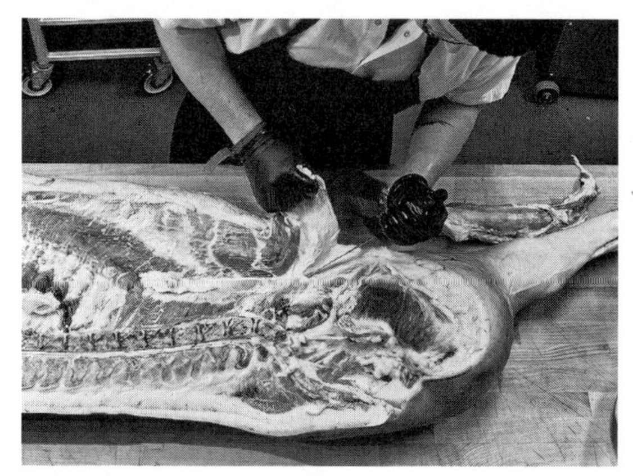
Preparing to remove the leg.

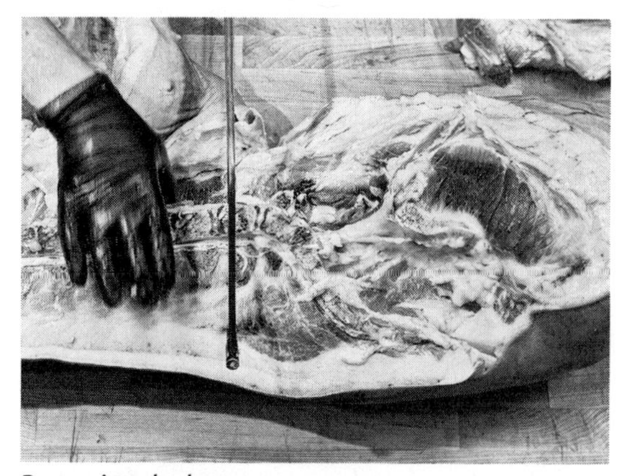
Removing the leg.

To remove the leg, saw through the spine, two vertebrae up from the tail. As soon as you saw through the bone, finish cutting through the meat with your breaking knife.

Next, Tyler will remove the sirloin from the leg, first by sawing through the hip, just above the aitchbone, and finishing with the knife. Don't worry if you miss the mark. Leaving a bit more sirloin on the ham, or vice versa, is not uncommon when you're starting out.

Now cut behind the pelvic bone to remove it from the sirloin, and if you aim to prepare tasso from the sirloin, take the skin off. Alternatively, you can make cutlets, marinated and pan braised as in pork banh mi sandwiches, or you can add sirloin meat to the grind.

The next task is to remove the aitchbone from the leg. Cut behind it with your boning knife, staying close to the bone and searching with your knife for the perfect separation point: the ball joint between the femur and the aitchbone.

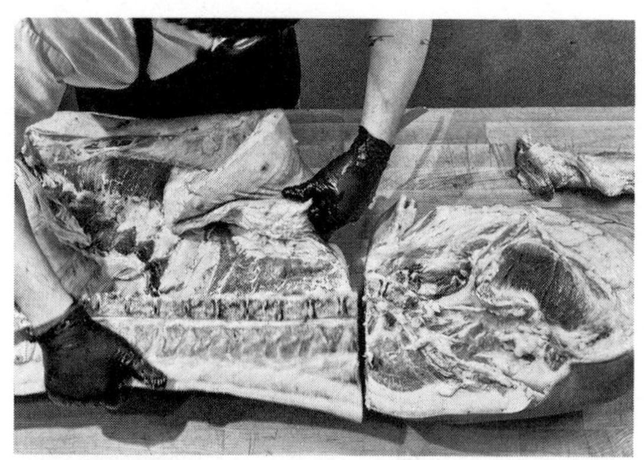
Removing the leg.

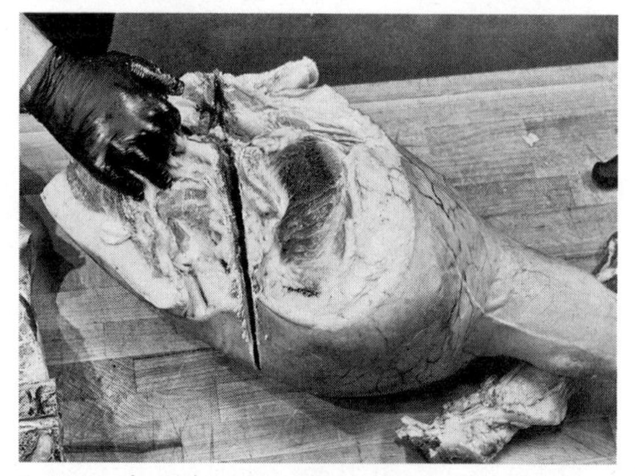
Remove the sirloin, just above the aitchbone. The pelvic bone will be on the sirloin. The aitchbone will remain on the ham. Both will be removed.

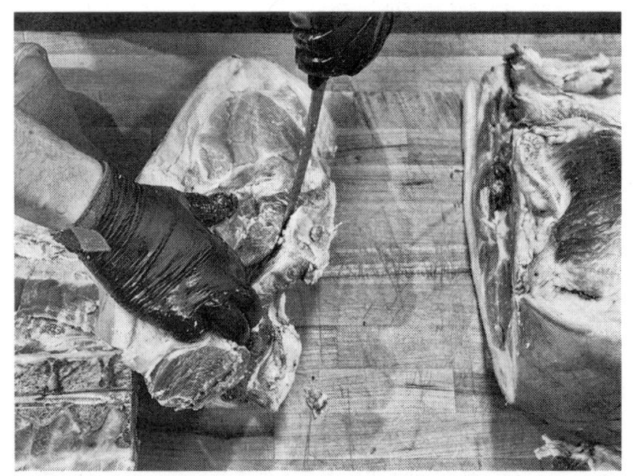

Deboning and skinning the sirloin for tasso.

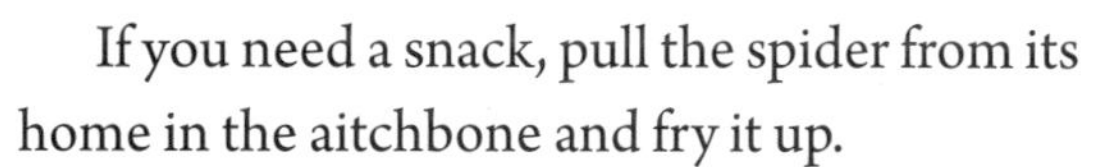

If you need a snack, pull the spider from its home in the aitchbone and fry it up.

At this point, the ham is ready to cure for prosciutto (p. 180).

You can also skip the prosciutto and break the leg for other projects. Tyler begins by skinning the ham, back to the top of the shank. You can see the end of the femur bone protruding from the top of the ham.

What we're going for next is the *culatello,*

Deboning and skinning the sirloin for tasso.

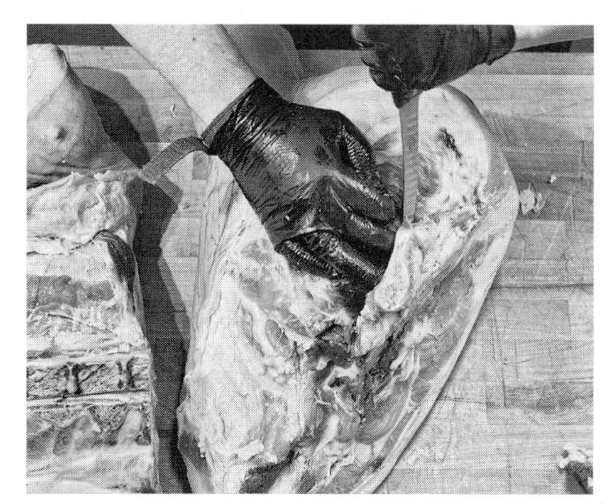

Removing the aitchbone.

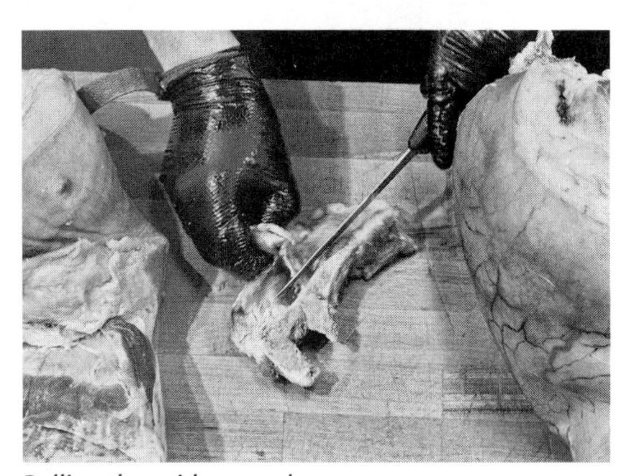

Pulling the spider muscle.

Skinning the ham.

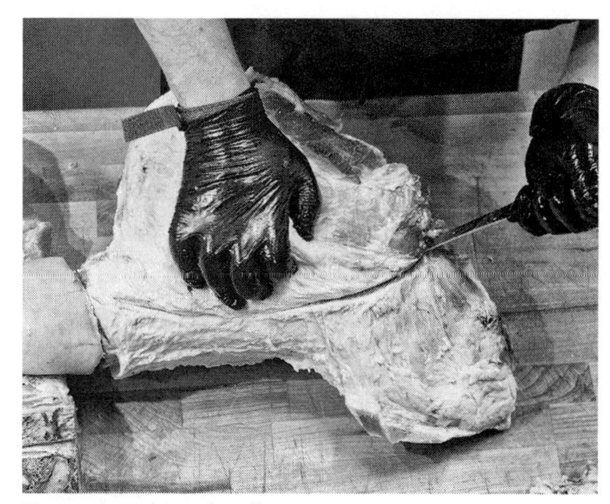
Removing the culatello.

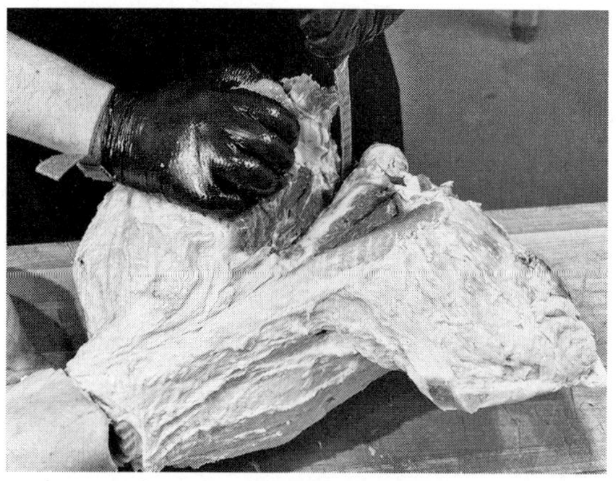
Removing the culatello. Notice the femur bone, and use it as your guide.

which is Italian for "small ham." This cut is a combination of the bottom round, top round and eye of round (in beef lingo), and sometimes the bottom sirloin ball tip. The culatello can be used in a multitude of preparations. In Italy the whole culatello might be cured with nitrites and sea salt and hung in a hog bladder in the curing chamber. If you don't have plans for it, you can always grind it for sausage or salami.

Tyler marks his cut, which rides alongside and atop the femur bone.

Next, follow the seam to remove the ball tip muscle from the culatello, if you so desire. You can add this meat to the grind or use it as a ham for curing.

If you want, you can further reduce the culatello by completely removing the bottom round muscle. This seam is easy to find, pretty much in the center valley of the culatello, once it is turned over.

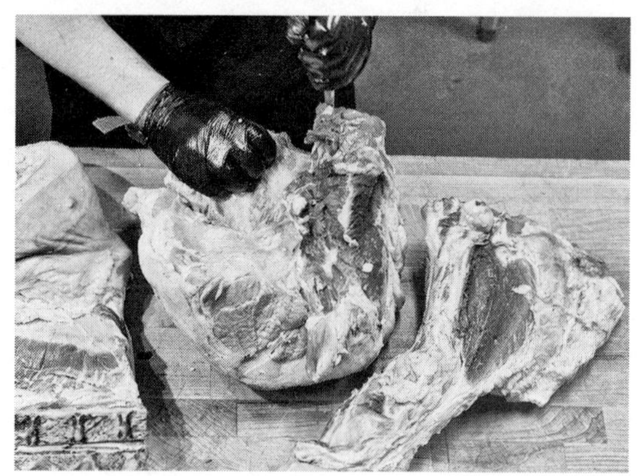
Seam out the ball tip muscle.

If desired, follow the seam to remove the bottom round.

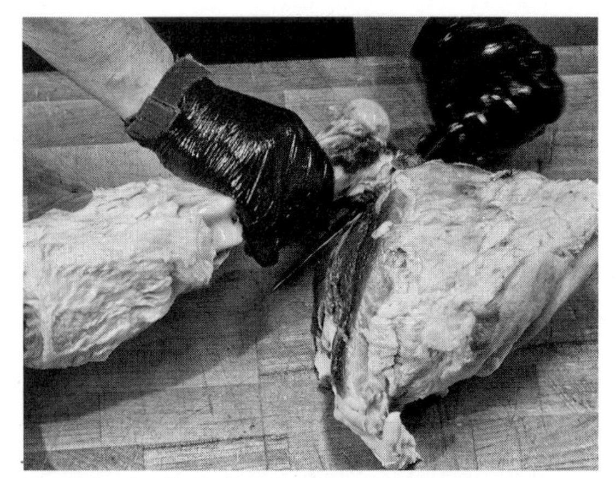

Remove the fiochetto from the leg bone.

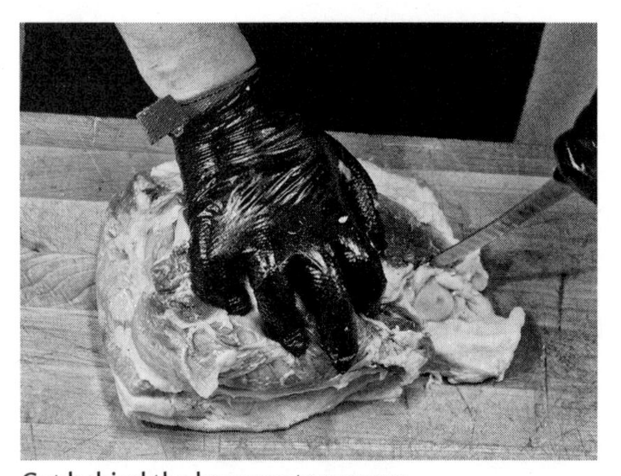

Cut behind the kneecap to remove.

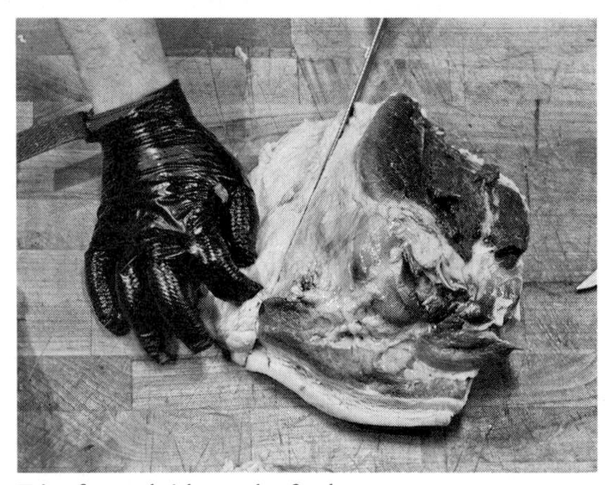

Trim fat and tidy up the fiochetto.

What's left on the femur bone is the muscle just behind the knee. On a beef, this is the "knuckle"; Italians call it the *fiochetto,* meaning "bowtie," a way of talking about pig I rather like. We'll remove this muscle by cutting close to the bone, and tie it up for the smoked fiochetto ham, page 182.

To prep it for brining and smoking, remove the kneecap, trim some (not all) of the fat, and tie.

Next, Tyler works to separate the shank from the trotter, which he does with his knife. Notice his technique of working with the joint movement to find the sweet spot. He bends the joint over the edge of his block to get his bearings.

Tyler plans to put shank meat in the grind, but you can also braise pork shanks, or crosscut them to produce pork "osso bucco." Trotters can be pickled, used for stock, or boiled and flavored to produce aspic for your next terrine.

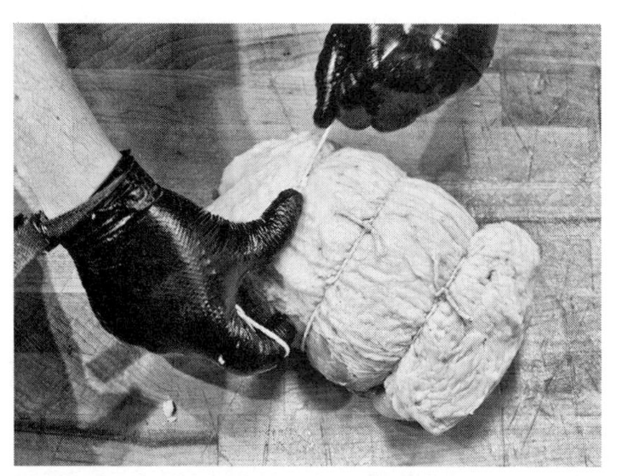

Tie up the fiochetto ham to ready it for curing and smoking.

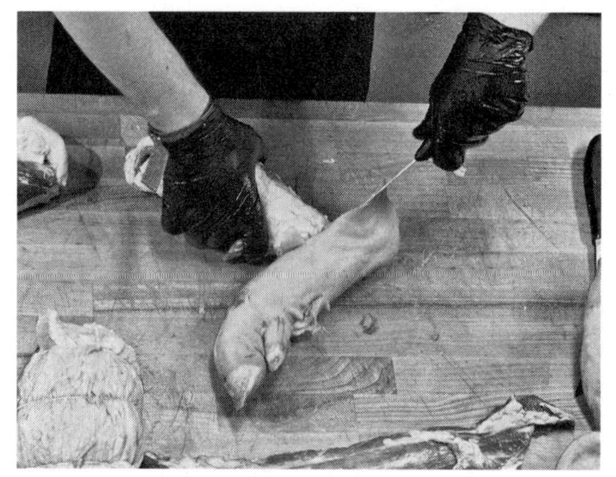

Tyler separates the shank and the trotter.

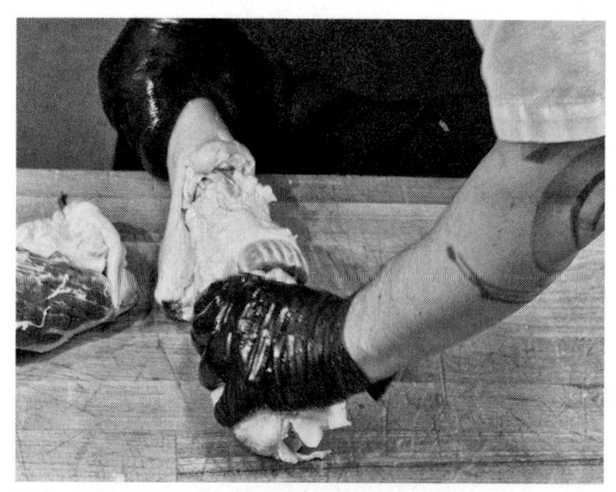

Tyler separates the shank and the trotter.

Next, Tyler counts ribs and marks his cut between the 4th and 5th ribs to distinguish the shoulder primal from the loin and rib section. He uses his knife to mark the cut, then saws through the ribs with the bone saw, finishing with the breaking knife.

We'll work with the loin and belly first. Tyler removes the flap meat: outside skirt, inside skirt and bavette. The skirt steaks are on top of valuable belly, so use care as you undercut them, and look for the seams to guide you.

Many people leave the flap meat cuts on and cure the belly for bacon, but in my opinion they create inconsistencies in the cured product. They are different muscles from the typical fat and muscle within the belly, and are very thin. This, coupled with their position on the carcass, will cause them to take in more salt and cure, so they dry out quickly during smoking. Best to remove them and give them more of a chance. Pork skirt steak is fantastic with lemon, butter and salt, fried quickly in

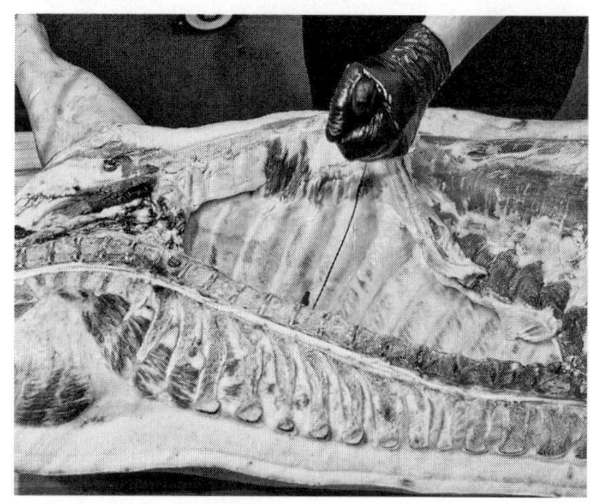

Removing the shoulder.

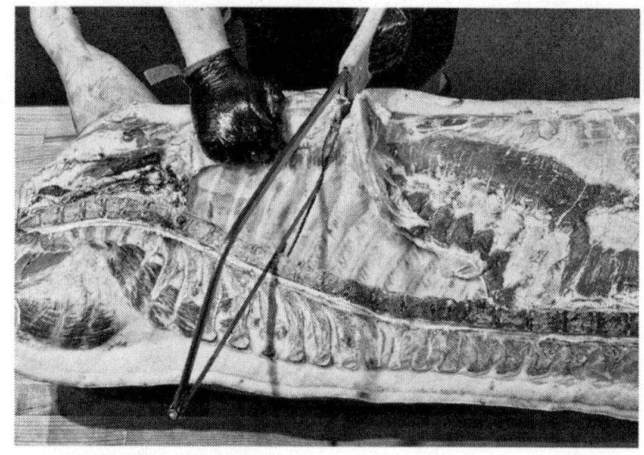

Removing the shoulder.

Tyler cuts away the outside skirt.

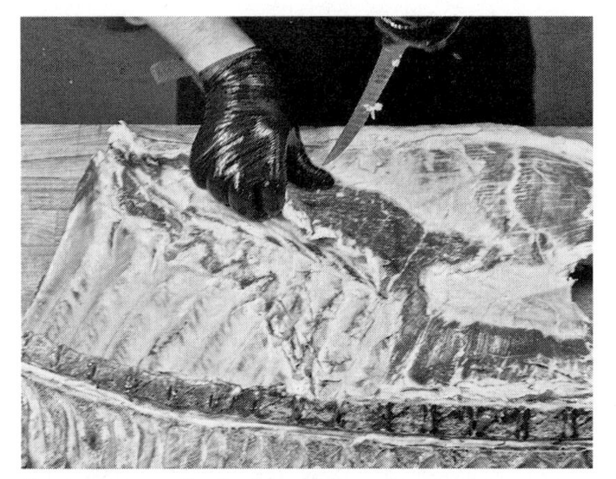

Next, remove the inside skirt.

a hot dry skillet. Bavette is wonderful butterflied, rubbed with salt garlic and chili paste, and grilled. Use your imagination, knowing that these cuts prefer a marinade and high-heat cooking.

Now it's time to separate the belly from the loin. Begin by using your boning knife to measure the eye of the loin, then add half of that distance to discover your cut mark.

Next, use the bone saw to begin cutting through the ribs. Stop when you get to the meat, and finish with your knife.

Pull the spare ribs from their home on the belly by undercutting them. You'll have to decide how much meat you want to leave on your ribs, and how much to leave for the belly.

Next, Tyler will separate the ribs from the strip loin. This cut is traditionally made between the 13th and 14th ribs, but Tyler moves up one, cutting between 12 and 13, because he knows we want to cut baby back ribs, and he's figuring a nice, eight-rib rack. This is a perfect

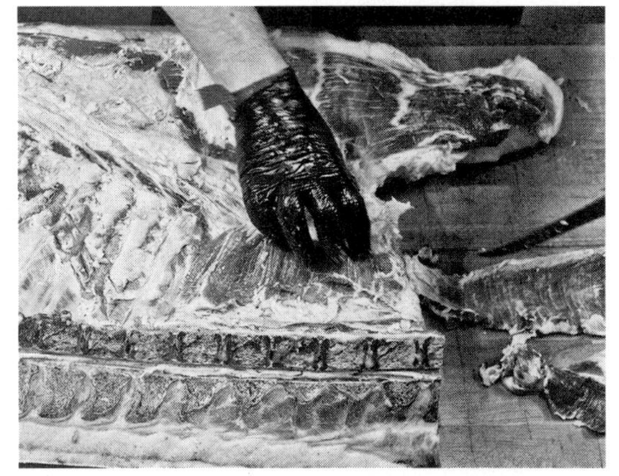

Tyler's hand on the bavette, ready to remove.

Team flap meat. From top: Inside skirt, outside skirt and bavette.

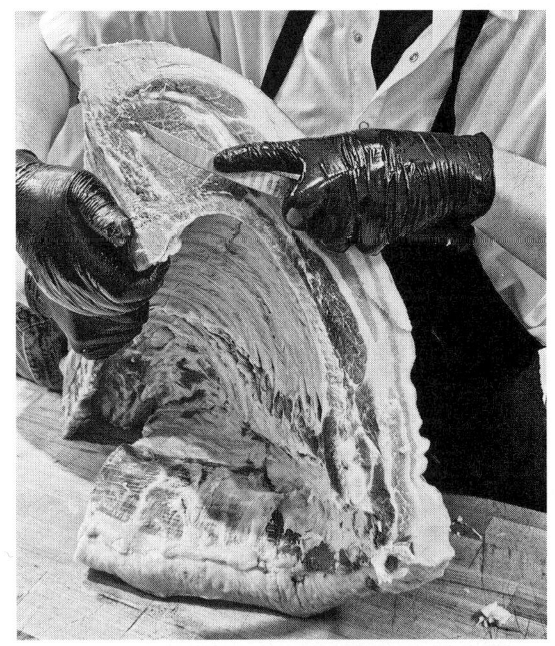
Tyler measures from the eye of the loin with this boning knife.

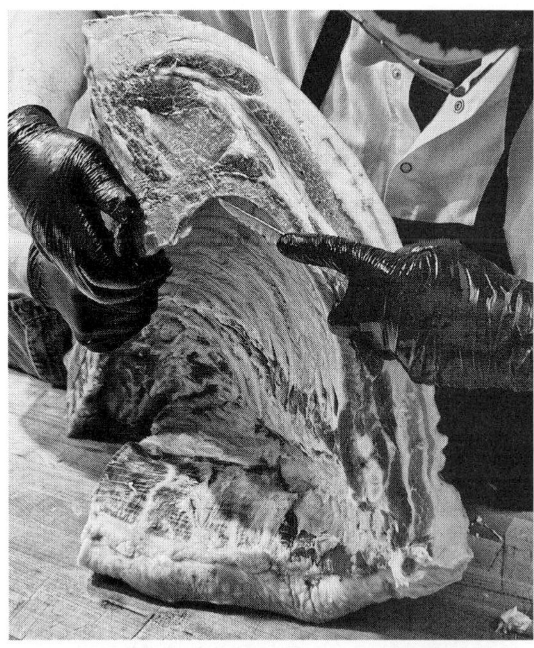
Use your mark to measure that same distance from the center of the loin to figure out where to cut.

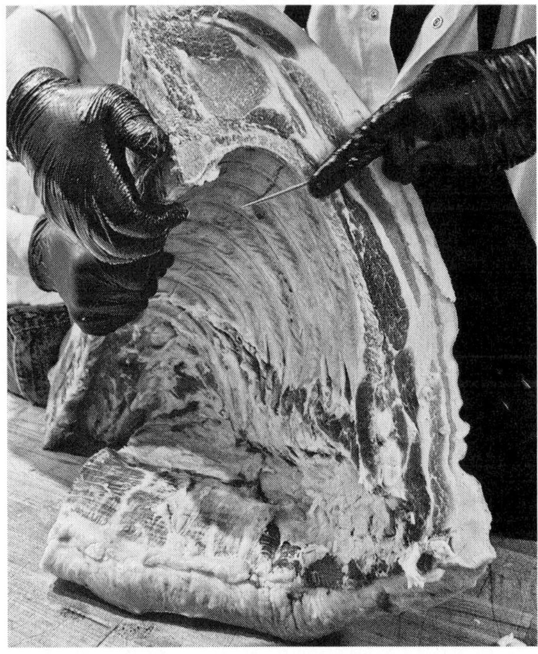
Mark your cut.

Use the saw to begin separating belly from loin.

Finish the cut with your breaking knife.

example of how you might want to modify decisions based on your needs.

To make this cut, mark and begin with your knife, using your saw to get through the bone.

We'll work with the rib first. Begin by turning the rib on end and sawing off the chine bone.

If you want to cut bone-in, skin-on pork chops, now's the time.

We are going to remove baby back ribs, instead of bone-in chops. Start by making a cut through the ribs to square off the rack, just below where the spine used to sit.

Now undercut the ribs, making them as meaty or as bony as you want. Now you can see why ribs are priced (and cooked) differently:

- Country ribs: from the shoulder; flavorful but tougher meat
- Spare ribs: from the belly; fattier with less meat

Removing the spare ribs.

Removing the spare ribs.

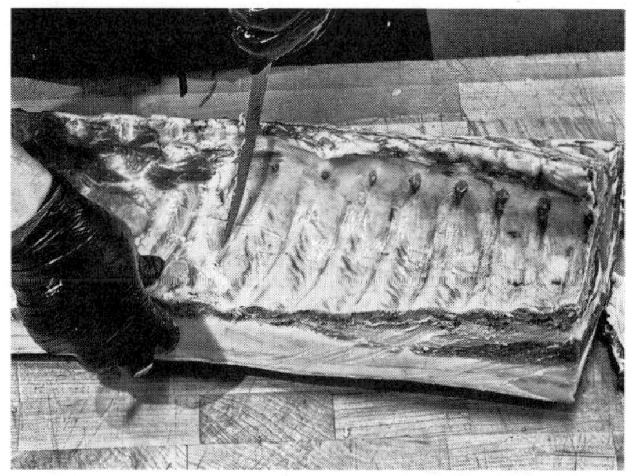

Separating the ribs from the strip loin. The rib portion is on the right, the strip loin on the left.

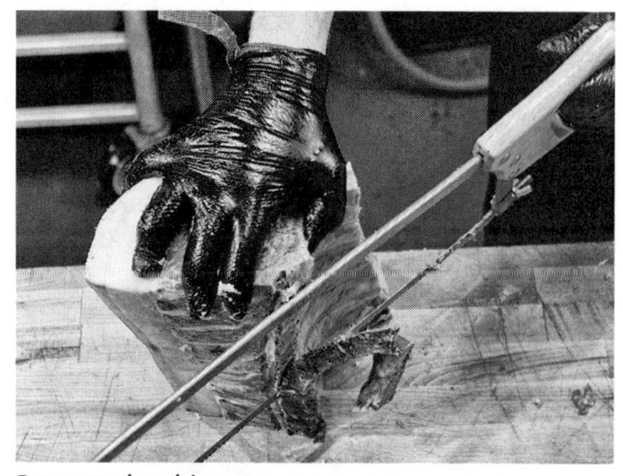

Remove the chine.

Preparing to cut bone-in, skin-on pork chops.

Square off the rack.

- Baby back ribs: from the prized loin; meaty and grill-worthy

You'll finish by removing the bones from the other side of the loin, too.

What's left is essentially the boneless loin, which you'll need to trim and skin before using as a whole loin roast. Or save it to put into your porchetta (more on that later).

Removing baby back ribs.

Finish pulling bones from the loin.

Deboning the loin.

Moving on to the strip loin, next, Tyler begins by boning it out. Use the bone to guide you, keeping your knife as close to the bone as possible.

Remember to save the bones for stock. Next, skin the boneless loin. Skin can be baked or smoked to make delicious treats for your pets, or baked until crispy and put into soups or beans. A thrifty and genius trick is to bake squares of skin until crispy, then buzz them up in your food processor. You can store the ground skins as they are, or combine them with salt and use them in spice blends or in breading for your favorite fried foods. You can also boil skin, to harvest its collagen proteins.

After skinning the loin, Tyler removes a layer of back fat, and then cuts boneless, skinless pork chops.

Let's finish up with the belly by producing a porchetta, which will combine our boneless loin from the rib with the belly to form a beautiful stuffed roast. (See porchetta with persimmon, chestnut and pine recipe, p. 154.)

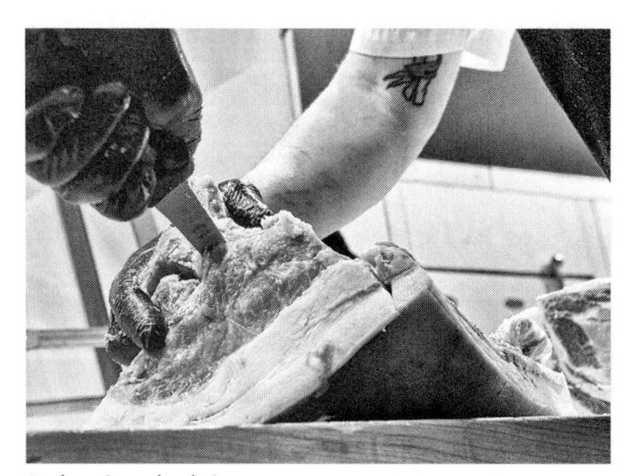
Deboning the loin.

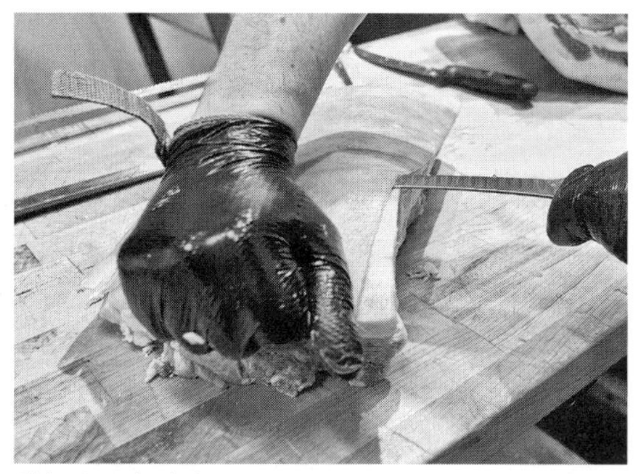
Skinning the loin.

Cutting boneless pork chops.

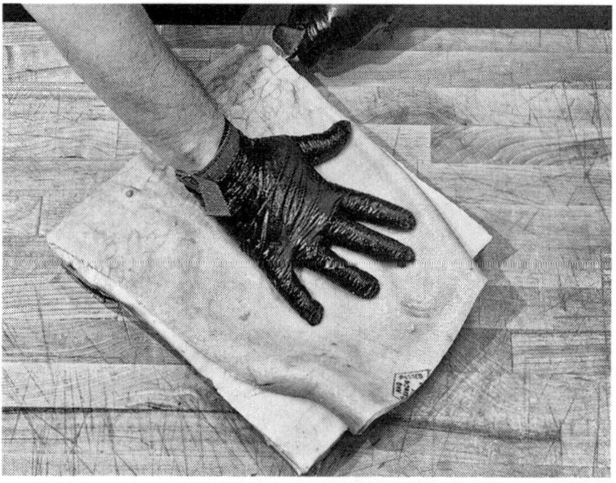

Pull the skin mostly off to begin bookending the belly portion.

Use the section of loin that you have to measure off the amount of belly you need. I suggest cutting the thinner side of the belly (farther from the shoulder) for porchetta. The meatier end of the belly that lived closer to the shoulder should be the portion you cure for bacon or pancetta, or use for braising later.

Square off the belly you identify for porchetta. Now you will "bookend," or spiral cut, the belly. To bookend, Tyler first skins the portion of belly just until he reaches one end, but stopping so that the skin is still attached along the edge.

Next, he'll cut halfway down into the belly meat and begin splitting in the opposite direction, stopping at the other end of the belly so that the meat is still attached along the edge. When finished, the belly should open like a trifold booklet.

The next step will be to go at the belly with a meat tenderizer, or cut slits in it with your boning knife. Prepare the brine or stuffing, per your recipe.

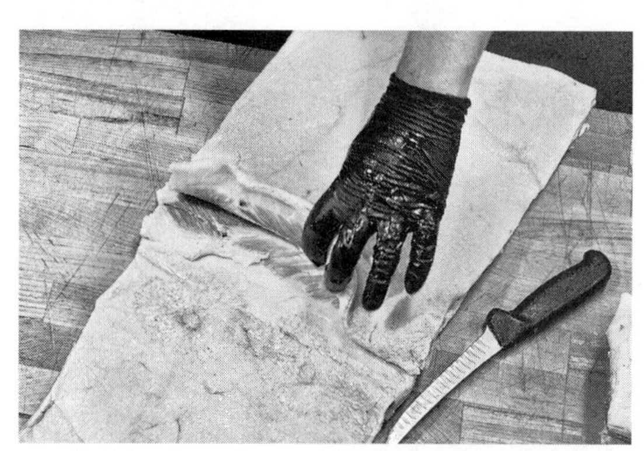

Notch down into the meat halfway, then begin to peel it back in the opposite direction from your previous cut.

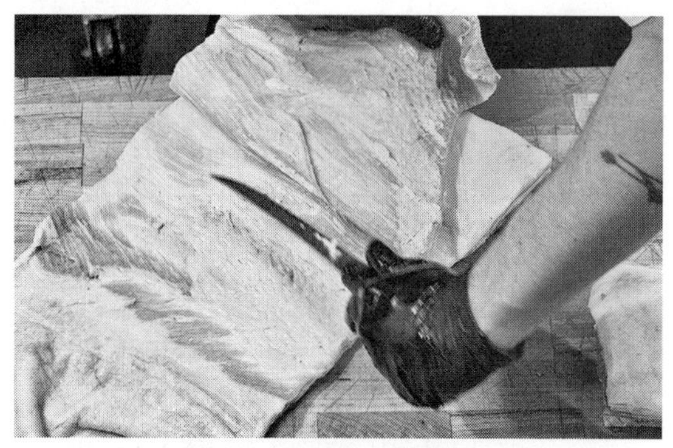

Continue cutting until you reach the end, leaving the meat attached along the edge.

The bookended pork belly.

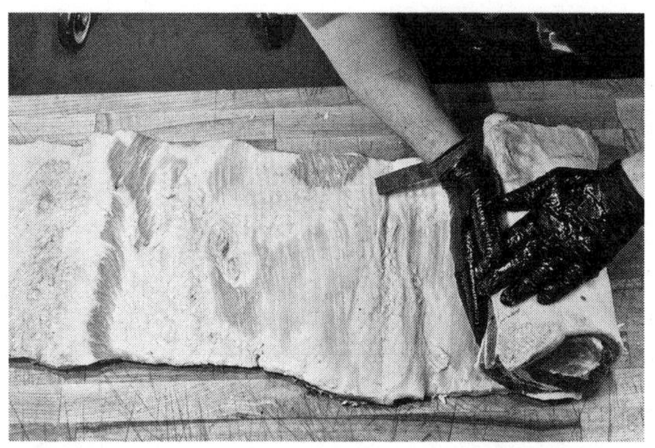

Roll the boneless, trimmed loin into the bookended belly.

The finished porchetta.

When you're ready to assemble the porchetta roast, place the boneless, trimmed loin onto the inside of the meat-end of the bookended belly. Begin to roll it up inside the belly, nice and even. When you're finished, the skin will be on the outside and the boneless loin will be wrapped in a spiral of scrumptious pork belly. Trim the ends and prepare as recipe directs. You'll tie it up after you stuff it per recipe instructions, but in our photo, it is already tied, just to give you the full effect.

Porchetta is also delicious smoked, which you should consider if you want an alternative to the traditional roasting. When you're done serving a porchetta roast for dinner, chill it, slice thinly, and use for sandwich meat.

Now it's time to deal with the pork shoulder. The first step is to separate the upper shoulder, or the Boston butt, from the lower shoulder, called the picnic.

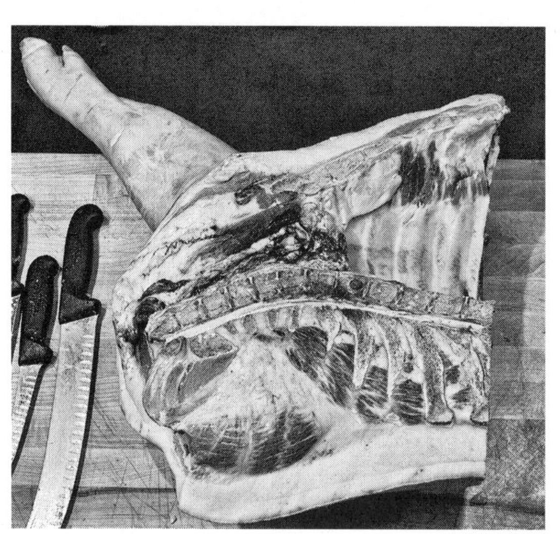

The pork shoulder primal.

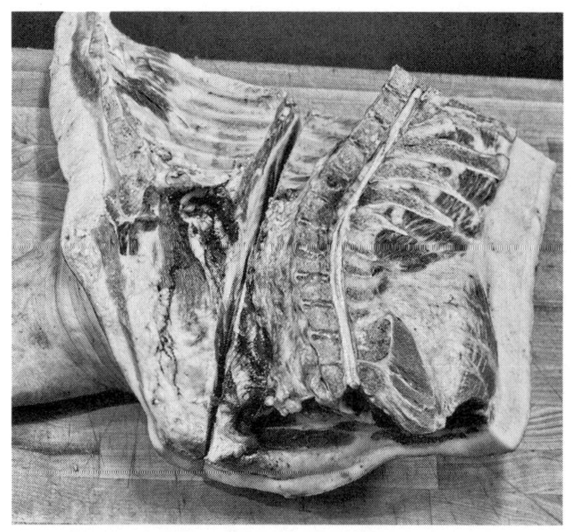

Separating the butt from the picnic.

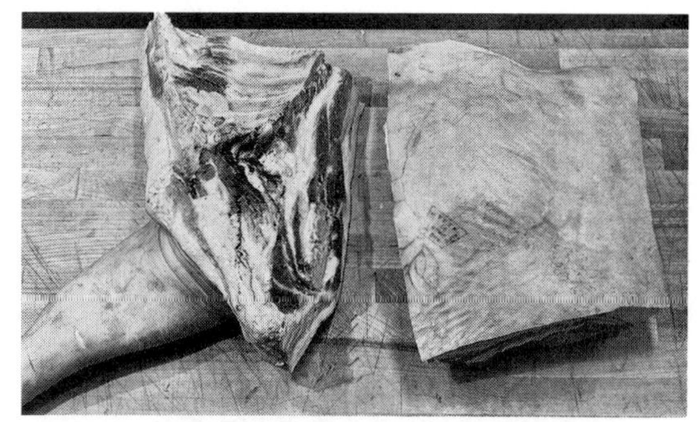

The picnic, at left, is home to the country ribs, which you can remove by undercutting them. We will leave them on for smoking. The Boston butt is pictured at right, skin side up.

Begin by placing the shoulder skin-side down on the bench and marking your cut parallel to the spine, below the lowest curve in the spine. You're essentially splitting the shoulder in half.

Use an alternating combination of saw and breaking knife to get through the ribs, then the meat below them, then the scapula, and then the meat below and the skin.

Next, undercut the chine and rib bones on the butt to remove them completely.

The butt is often smoked whole, bone in, or boned out and used for sausage. Here, we will isolate the "collar," which is back muscle comprised of some of the best-hidden, most tender muscles in the carcass. This is often the cut you'll see sold as a boneless pork shoulder roast. The collar is also a prized cut in the barbeque scene. Tyler likes to cure it whole and then warm smoke it for his meat case. He calls

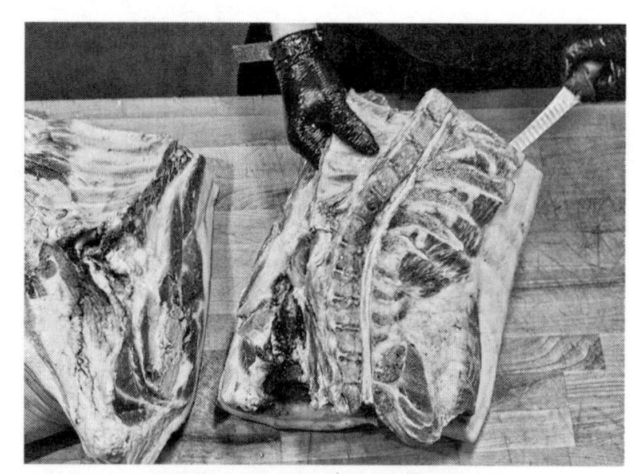

Removing chine and rib bones from the Boston butt.

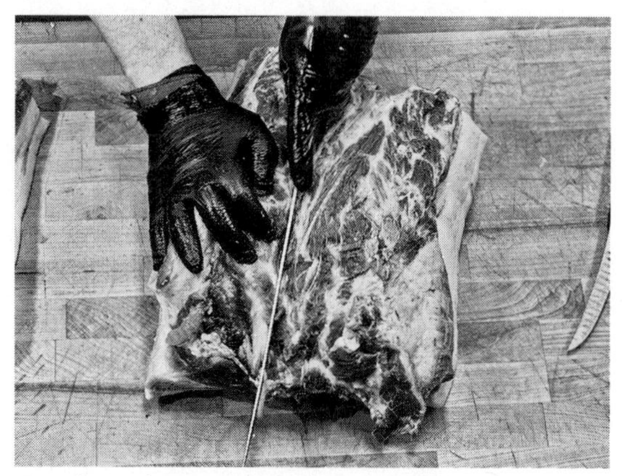

Cut along the seam to isolate the collar.

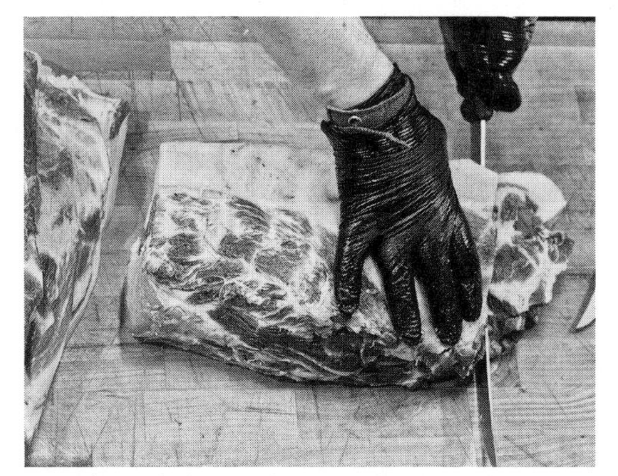
Trim the ends for good presentation.

The collar, on right.

this "hot smoked capicola," and slices it for sandwich meat. Delicious.

Remove the collar with your breaking knife, right along the seam. You can easily differentiate the larger collar from the thinner muscles beside it.

The remaining portion of the butt can likewise be cured, smoked, roasted or trimmed for sausage. Just for the fun of it, Tyler will demonstrate removing the pork flat iron (you'll recall this is a specialty steak cut from the beef carcass, within the shoulder clod subprimal).

First, remove the scapula by carefully cutting around it to pull it out completely.

Once you've pulled the scapula, you'll notice a rectangular-ish muscle, which sat on top of the shoulder blade. Seam this out, and you'll have your pork flat iron.

The good news is that the pork flat iron doesn't need to be trimmed out like the beef flat iron. It can be grilled or pan cooked as it is.

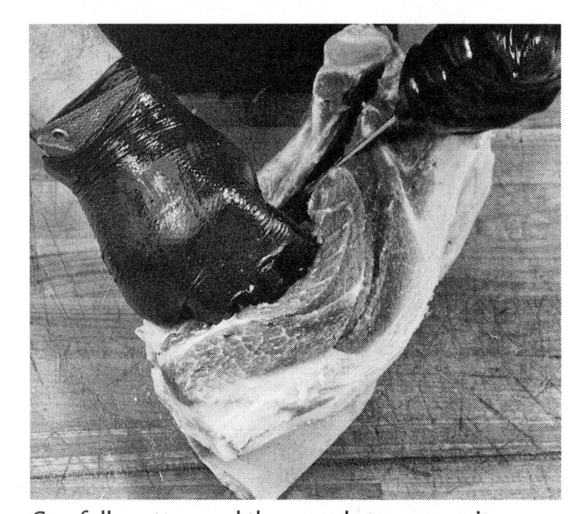
Carefully cut around the scapula to remove it.

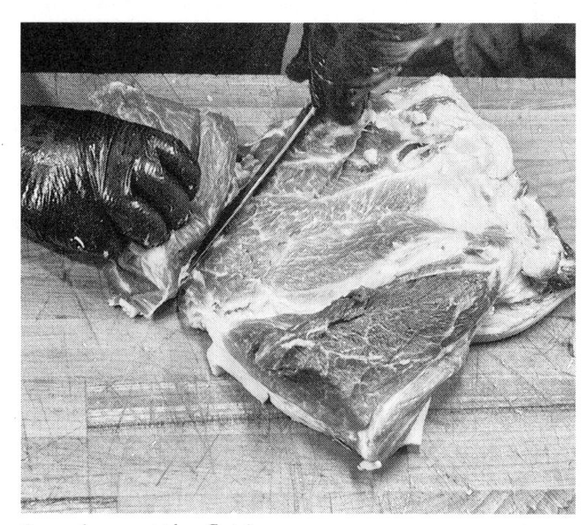
Seaming out the flat iron.

At this point, you'll have some additional muscles, and fat from the butt that you can contribute to your sausage trim.

Finally, prepare the picnic for smoking. First, remove the trotter. Find the joint at the ankle, cut around the bone with your knife, then saw through the leg to pull the trotter completely.

Next, work to mostly skin the picnic, carefully cutting so that you leave as much fat on the meat as possible. The object is to remove the skin to a point on the shank, then score the fat so you can rub your spice or cure mixture in, and then replace the skin before smoking. The skin is important for holding in moisture during cooking but will inhibit the binding of spice or cure to the picnic; thus it is crafty to be able to move it out of the way when seasoning.

The pork flat iron, directly left of Tyler's cimeter.

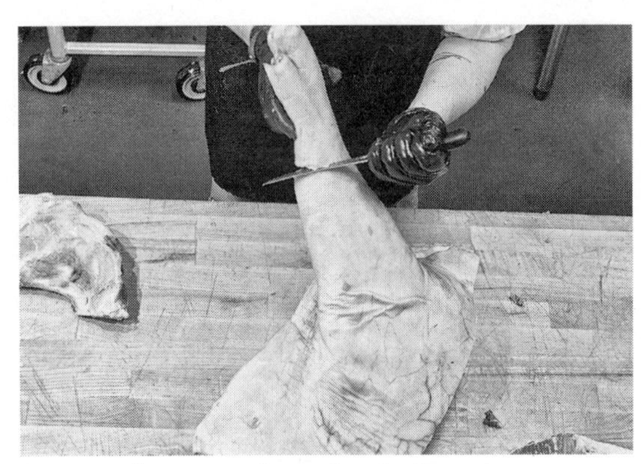

Remove the trotter from the picnic.

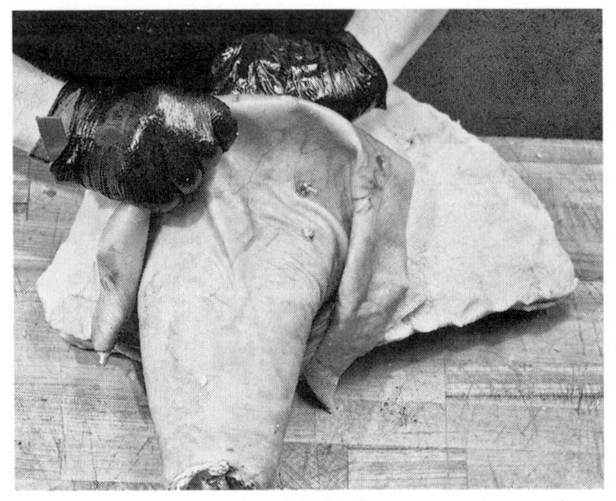

Partially removing the skin before seasoning and smoking the picnic.

Cooking With Pork

I could sit here trying to think of all the adjectives in the world to describe pork cooked right. But I'm having trouble, and the truth of it is that well-executed pork is so good it's almost a feeling. It is beyond taste. This, I reason, excuses me somewhat from having to describe it.

Here I go, though, just to say that anyone who's had pork done right has had a nose of the steam much like that off of warmed, ripe fruit, and can forever recall that first moment of sweet and smoky fat all over his lips. Pork calls for vinegar, and salt, and earthy, crunchy, herby, holy foodstuffs of many kinds. And fire. OK, whatever you do with it will be just fine. That's the magic of the pig.

The recipes here are for fresh pork, and don't involve curing. You'll find plenty more recipes in the Charcuterie chapter, which covers sausages, smoking, curing and fermenting.

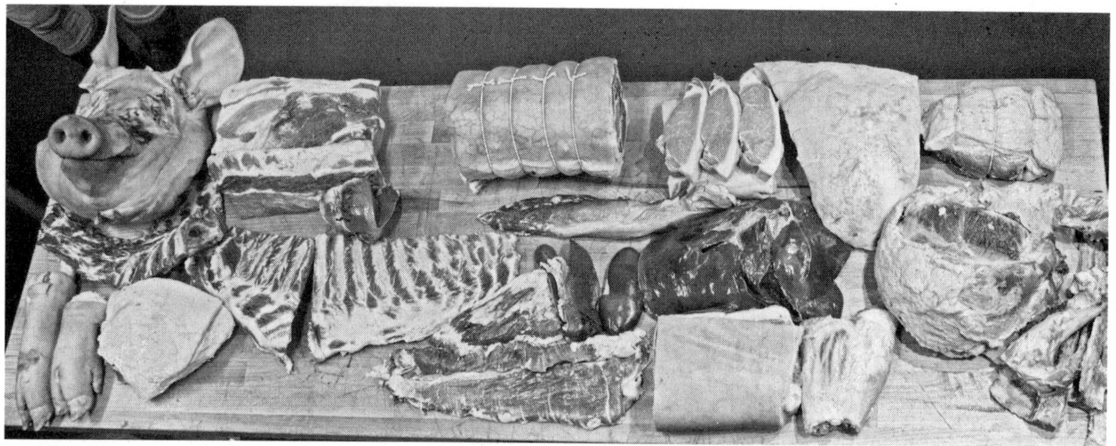

The whole hog. Top left, clockwise: the head, the collar (with pork heart just below it), our porchetta (whole tenderloin just below it), boneless skinless loin chops, sirloin, fiochetto, culatello (with trim and bones beside it), shank, folded belly (liver, kidney and spleen just above it), flap meat, spare ribs, country ribs, trotters and baby back ribs.

Pulled Pork with Hot Vinegar Sauce, Chow Chow and Corn Pancakes

FOR THE PORK

One whole picnic, bone in and skin on
½ cup brown sugar
4 Tbsp kosher salt

FOR THE VINEGAR SAUCE

3 cups apple cider vinegar
1 cup water
¼ cup brown sugar
2 Tbsp paprika
2 Tbsp black pepper
2 tsp kosher salt
2 Tbsp red pepper flake

FOR THE CHOW CHOW

3 red bell peppers, seeded and chopped
1 yellow bell pepper, seeded and chopped
1 green bell pepper, seeded and chopped
1 green tomato, chopped
1–2 jalapenos, seeded and minced
2 sweet onions, chopped
½ head of cabbage, chopped
¼ cup kosher salt
2 cups evaporated cane juice crystals
2 cups distilled vinegar
1 cup water
1 Tbsp mustard seed
2 tsp celery seeds
1 tsp turmeric
½ tsp fresh ground black pepper

FOR THE CORN PANCAKES

½ cup flour
½ cup cornmeal
1 Tbsp sugar
1.4 tsp baking soda
¼ tsp salt
1¼ cups fresh sweet corn
⅔ cup buttermilk
2 Tbsp butter, melted
1 egg

Combine the brown sugar and the salt and rub the picnic all over. Refrigerate overnight.

Combine all the chow chow veggies and the salt in a nonreactive mixing bowl. Refrigerate overnight.

The next morning, prepare the smoker. Rinse the picnic, pat it dry and smoke it at 150°F until the bones are loose and the meat is tender and juicy.

While the meat is smoking, remove the chow chow veggies from the fridge, drain and rinse. Transfer to a deep cast-iron pan or Dutch oven and add the sugar, vinegar, water and spices. Combine and bring to a boil, then reduce heat and simmer for 5 minutes. Then pack the hot chow chow into quart mason jars, lid them and cool completely. Refrigerate.

Just before serving, mix all vinegar sauce ingredients and heat thoroughly, until the sugar and salt dissolve. Pull the pork and sauce it, and keep warm while you make the corn pancakes.

In a mixing bowl, combine the dry ingredients. In a separate bowl, combine the corn, buttermilk, butter and egg, and blend with an immersion blender until pureed. Add the puree to the dry ingredients and stir until just combined. Heat a cast-iron skillet over medium heat, grease with butter or bacon fat, and add ¼ cup batter per pancake, griddle cooking on both sides until golden brown.

Serve two corn pancakes, smothered with the sauced pulled pork and topped with the colorful chow chow.

Pork Banh-Mi Sandwiches with Quick Pickles

Colonialism was nasty. I think we could have figured this one out without such nonsense, but here it is, nonetheless: the banh mi sandwich, a fusion of French food and Vietnamese flavors. You can be thrifty and traditional, and make this sandwich with ground pork or pork trim and scrap, but it's made here with pounded cutlets.

FOR PORK

2 lb. pork sirloin or tenderloin
¼ cup fish sauce
3 Tbsp rice vinegar
3 Tbsp brown sugar
3 Tbsp tamari
1 tsp sesame oil
3 garlic gloves, minced
1-inch piece of ginger, minced
1 bunch chives, minced
1 tsp black pepper, fresh ground
2 serrano chilies, crushed
2 Tbsp neutral vegetable oil
Kosher salt

FOR QUICK PICKLES

½ pound mixed julienned root vegetables, such as carrot, radish, beet, daikon, etc.
½ cup water
1 cup distilled vinegar
1 tsp kosher salt
2 Tbsp evaporated cane juice crystals

FOR SANDWICHES

2 jalapenos, sliced
¼ cup liver pâté (see recipe p. 173)
Homemade aioli or mayonnaise (use aioli recipe on p. 116)
1 bunch cilantro
4 6-inch lengths of good, fresh, ciabatta bread

Slice the pork into medallions or cutlets. Place the cutlets between waxed paper and pound with a rolling pin until tender and flat. Mix all marinade ingredients, except for oil and salt. Put meat into marinade and refrigerate, covered, at least three hours, and as much as overnight.

Prepare the pickles. Place veggies into a nonreactive bowl or mason jars. Bring water, vinegar, salt and sugar to a boil in a small saucepan to create a brine. Pour brine over julienned vegetables and cool, then transfer to a refrigerator.

When ready to assemble sandwiches, remove meat from marinade, reserving marinade. Warm oil in a cast-iron skillet and begin cooking the pork in batches, pouring some of the reserved marinade over it and allowing it to caramelize while the meat cooks through. Transfer cooked meat to a plate and sprinkle with kosher salt.

To assemble the sandwiches, slice the ciabatta loaves lengthwise and lightly toast them, dry, in the oven. Smear one side with aioli and the other with pâté. Fill with the meat, jalapeno, quick pickles and cilantro. Serve immediately.

Braised Pork Ribs with Rooster Sauce and Balsamic Vinegar

Homemade rooster sauce is something to have in your fridge at all times. It's great on eggs in the morning, and it's great for that stir-fry you made out of all your leftovers. This recipe for it makes two cups, but you'll only need a quarter of that for these braised ribs.

FOR RIB RUB

- 2 slabs pork ribs (country, spare or baby back, or a mix), cut plenty meaty
- 3 Tbsp kosher salt
- 3 Tbsp brown sugar
- 1 Tbsp chili powder
- ½ tsp onion powder
- 2 garlic cloves, minced
- Pinch of cayenne pepper

FOR BRAISE

- ½ cup rooster sauce
- ½ cup balsamic vinegar
- 1 Tbsp honey

FOR ROOSTER SAUCE (MAKES 2 CUPS)

- 1 lb. jalapeno peppers, seeded and sliced
- ½ lb. serrano peppers, seeded and sliced
- ½ lb. sweet red bell peppers, seeded and sliced
- 6 cloves garlic, minced
- 6 Tbsp evaporated cane juice crystals
- 1½ Tbsp kosher salt
- ½ cup distilled vinegar
- ½ tsp xanthan gum

To make rooster sauce: Combine the peppers, garlic, salt and cane juice crystals in a glass jar with an airlock. Ferment at room temperature for a minimum of 5 days. Experiment with longer fermentation to create a stronger flavor. When you're satisfied with the ferment, transfer everything to a blender and add the vinegar and xanthan gum, and puree. Strain through a chinois or cheesecloth and store in a jar in the fridge.

For the ribs, mix up your rub and rub the ribs the night before. You'll want to wrap them up tightly once you've rubbed them and chill them overnight.

When you're ready to cook the ribs, remove them from their wrappers and place them in a roasting pan. Preheat the oven to 250°F. In a small saucepan, combine the braise ingredients and heat through. Pour half of the glaze into the roasting pan and toss the ribs to coat. Cover with foil and place in the preheated oven for 2 hours.

After the ribs braise, remove the foil and baste them with some of the remaining glaze. Turn the broiler on in the oven and brown the ribs, caramelizing the glaze. Remove from oven and toss with remaining glaze before serving.

Chicharron with Apple Butter and Cilantro Crème Fraîche

Chicharron is pork skin, and my favorite way to cook it comes from the Domenican Republic—with plenty of meat left on. Some folks make it with the skin only. Either way, the essential process is to confit the pork belly and deep fry it at the end. What could possibly be better?

FOR THE CHICHARRON

2½ lb. pork belly, skin on, sliced into strips
Baking soda

APPLE BUTTER

See "Sundries," at the end of this chapter

FOR CRÈME FRAÎCHE

1 pint heavy cream
2–3 Tbsp cultured buttermilk or whey
Pint mason jar
Salt
Chopped fresh cilantro

To prepare the chicharron, rub the pieces of belly with baking soda, place in a nonreactive dish, and let sit overnight in the fridge. (This step is optional, but makes a better, crunchier snack.) Next, prepare the crème fraîche by pouring the heavy cream and buttermilk or whey into the jar. Stir, place a lid on the jar, and leave on the counter overnight.

The next day, take the belly out and place it in a large, deep, cast-iron skillet. Put a half inch or so of cold water in the bottom and heat over medium-low heat. The fat will begin to melt off of the belly, mingle with the water and slowly cook the meat. You can stir the meat now and again, but this is not necessary. Keep an eye on it, and when you're sure all the fat has cooked off, turn up the heat. This is the essential step that makes chicharron—crispy-frying it in its own fat. You may want to have a loose-fitting lid or splatter-screen on hand, for when it gets really serious.

Chop the cilantro and add it, with the salt, to the crème fraîche. Refrigerate.

Make the apple butter per instructions in "Sundries," at the end of this section.

Serve the hot chicharron on a platter, with hot apple butter and cold crème fraîche for dipping.

Lard

Best to always have lard in your fridge, in case you ever run out of butter or chapstick, or want to fry egg rolls or make perfect pie dough. There is lard, and then there is leaf lard. Lard comes from fatback, the thick, firm fat on the hog's back. Leaf lard is the fat from around the organs. Leaf lard is traditionally rendered separately for pastry making, as it has a finer, cleaner quality that lends to flakiness and a more neutral flavor. Rendered back fat is best for frying. In my house, back fat is rarely rendered at all, because it is the firm fat needed for making sausages, rillettes and lardo. With so many worthy projects, it gets used up quickly.

4 lb. pork fat
1 cup water (optional)
slow cooker

Cube up the fat as small as you care. The more surface area, the more lard you'll get. Some people go so far as to send the fat through the course plate of the meat grinder. I usually just chop it up with my knife into 1- to 2-inch cubes. Hook up an extension cord and place the slow cooker on the porch, so you don't stink up your house. Pour the water into the bottom of your cooker and then put all the cubed fat inside. Turn the slow cooker on low and get on with your day. You'll want to stir the fat now and again and keep an eye on it. If it cooks too long or too hot, you'll degrade the quality of the lard. The key to rendering is low and slow. You'll know it's done when the chunks have melted down as far as they can.

Strain the lard through cheesecloth or a fine sieve. Discard the contents of the sieve and cool the strained lard to room temperature, in a nonreactive container with a lid. Transfer to the refrigerator or a cool cellar for storage.

Basic Pie Crust

1¼ cup all purpose flour
¼ tsp salt
3 Tbsp cold rendered leaf lard (above)
3 Tbsp cold cultured unsalted butter
Approx. 3–5 Tbsp ice-cold water

Combine the flour and the salt, then cut in the lard and butter with your fingers. Add ice water slowly, until the dough just holds together. A great way to hydrate flour for pastry dough (where you don't want too much water) is with a spray bottle. Once the dough holds together, wrap it in plastic. Chill in the fridge for 30 minutes before rolling on a floured board.

Pork and Pickle Pie

When I first started brainstorming the pork recipes for this book, I came up with some pretty solid ideas right off the bat, and felt attached to them. But then I got to the point where I realized I had too many recipes, and there were random ingredients that kept popping into my head that I simply had *to put in a recipe. The only logical thing was to put all of them together, betwixt pastry dough. That's right. Pie. This is a fabulous thrifty recipe and can be used to deal with trimmings, producing a hearty meal.*

1 lb. pork trim, finely diced, or 1 lb. unseasoned ground pork
1 large sweet onion, chopped
2 cloves garlic, minced
1 Tbsp butter
1 cup bread and butter pickles (see "Sundries"), chopped
½ cup chopped pecans
¼ cup chopped fresh parsley
Kosher salt
Fresh ground pepper
6 oz. blue cheese
2 basic pie crusts (p. 151)

Preheat the oven to 350°F. In a heavy cast-iron skillet, melt the butter over medium-high heat and add the meat, onion and garlic. After the meat browns, turn heat to low and dump in chopped pickles, pecans and parsley. Add salt and pepper to taste.

Roll out one pie crust and place it in a deep-dish pie pan. Fill the crust with the pork filling, add blue cheese, and then top with the second pie crust. Pinch the edges, prick the top, and brush with an egg wash of 1 egg beaten with 1 Tbsp milk. Bake for 40–45 minutes, until the crust is browned. Remove from the oven and slide a pat of butter over top of the browned crust. Serve with a worthy green vegetable, like marinated broccoli raab with lemon.

Breakfast Scrapple with Arugula, Eggs and Maple Syrup

Ultimate meat frugality is pronounced "Scrapple," a food of Pennsylvania Dutch origin. Variations are endless.

- 8 oz. pork heart, liver, or a combo of the two
- 2 lb. of pork trimmings, bones included
- 1½ quarts cold water
- ¾ cup bacon ends, roughly chopped
- 1 Tbsp maple syrup
- 1½ cups cornmeal
- ½ cup flour
- 1 tsp sage leaves, chopped
- 1 tsp thyme leaves
- 2 Tbsp kosher salt
- 1 Tbsp black pepper
- Half of a brown paper grocery sack, without printing on it
- 2 Tbsp butter
- Handful of flour for dredging
- ½ lb. bunch arugula
- 6–8 eggs
- Warm maple syrup

Lightly oil a 9-inch loaf pan and line it with plastic wrap so the wrap hangs over the edges on all sides.

Place the liver and heart into a small pot of water and simmer until cooked.

Place the pork trim and bones into a stockpot with the 1½ quarts cold water and bring to a boil. Lower heat, cover and simmer until the meat falls from the bones and you've produced a hearty stock. Strain, reserving liquid. Remove bones from the strained meat and trim and discard. Chop the meat and trim and combine with the bacon ends. Chop the cooked heart and liver and add to the bacon and trim. Add herbs, salt, pepper and 1 Tbsp maple syrup, and then combine in the stockpot with the reserved stock from earlier. Return the pot to the stove and bring to a boil.

Combine ½ cup flour and the cornmeal. As soon as the pot boils, whisk in the cornmeal and flour mixture, stirring until well combined. Reduce the heat to low, place the paper bag over the pot, and put the lid on top of it. Let thicken for 30 minutes.

After scrapple has thickened, remove the pot's lid and the paper bag and pour the scrapple into the prepared loaf pan. Cool to room temperature, then chill overnight in the fridge.

The next morning, remove the scrapple from the loaf pan using the plastic wrap edges you left hanging over the pan. Slice the scrapple loaf into half-inch pieces and dredge in flour.

Heat a small, cast-iron pan, then add the butter. Dredge the scrapple slices in the flour and fry them in the butter until golden. Remove fried, heated scrapple to a serving platter and drizzle with warm maple syrup. Keep warm while you fry the eggs soft and wilt the arugula. Serve scrapple with greens and eggs on top, and a little homemade hot sauce.

Porchetta with Persimmon, Chestnut and Pine

Paying respects to the forest's edge in Appalachia, where our pigs thrived, I developed this recipe to mimic the fragrant diet enjoyed by the hogs themselves. A good recipe for a special occasion, this one takes time and love.

FOR THE BRINE

1 gallon water
2 cups kosher salt
¼ cup black peppercorns
¼ cup evaporated cane juice crystals
3 cups green pine needles
2 Tbsp juniper berries

FOR THE PORCHETTA

5–6 lb. pork belly, skin on, bookended
2–3 lb. pork loin, boneless
6–8 ripe persimmons
3 Tbsp unsalted butter
3 Tbsp brown sugar
1 tsp nutmeg, freshly grated
3 tsp kosher salt, divided
3 cloves garlic, roasted in ⅓ cup extra virgin olive oil (reserve oil)
1 Tbsp fresh sage leaves
1 Tbsp fresh ground red pepper
1 cup chestnuts, roasted per instructions below, and roughly chopped

Two days before roasting, prepare the brine. Bring to a boil all the brine ingredients, then pull the pot from the heat and cool to room temperature. Then place the flat, bookended belly and the boneless loin into the brine and weight them so they are completely submerged. Refrigerate overnight.

The next day, remove the meat from the brine, rinse and pat dry. Score or poke holes in the pork belly skin, then flip the belly over and lightly score the belly meat. Mix the sage, red pepper, 2 tsp of the salt, the roasted garlic and the garlic oil. Rub generously over the belly and loin, inside and out. Place both pieces of meat into the refrigerator, not touching one another, and leave uncovered overnight.

The next day, roast the chestnuts by scoring them with a sharp knife and placing them on a baking sheet in a 350°F oven for 15–20 minutes. Cool them, peel them and chop coarsely. Set aside.

Trim and de-seed the persimmons. In a cast-iron skillet, melt the butter and add to it the brown sugar, nutmeg, sage and persimmons. Heat gently until the persimmons cook and release their juices and the aromas of all ingredients begin to meld.

Remove the meat from the refrigerator. Place the belly on a roasting pan and rub the persimmon mixture generously over the inside. Add the chopped chestnuts in a layer on top of the meat. Place the loin at one end of the belly and rub it with the persimmon mixture too. Roll the loin inside the belly and tie with butcher's twine every 2 inches. Allow the roast to rest at room temperature for 1 hour.

Preheat the oven to 500°F. Roast the porchetta for 30–40 minutes, then reduce the heat to 300°F to finish cooking, until a meat thermometer inserted into the center of the roast comes to 145°F. Let the porchetta rest at least 15 minutes after removing it from the oven, then slice it into quarter-inch rounds and serve with something like roasted root vegetables and a dandelion and wild mushroom salad.

Sauces and Sundries for Pork

Barbeque Sauce

One of many, many worthy recipes.

3 Tbsp bacon grease
2-inch piece of salt pork
2 onions, diced
8 garlic cloves, minced
2 cups brown sugar
2 lb. roma tomatoes, halved, salted, and baked at 300 for an hour or more
1 cup Dijon mustard
1½ cups pork stock
2 cups apple cider vinegar
1 Tbsp cumin
1½ cups bourbon
2 Tbsp sea salt
1 Tbsp fresh ground black pepper
cayenne, as much as you like

Heat the bacon fat in a large soup pot and sauté the onion and garlic. Add the remaining ingredients, in order, and simmer at least 1 hour. Blend with an immersion blender and simmer some more. Makes 1 quart.

Sauerkraut

This is true, traditional kraut. Everyone loves it. No one believes it is just cabbage and salt. Eat it with orange and garlic brats, page 169.

2 lb. cabbage, as fresh as possible
1 Tbsp + 1 tsp sea salt
Quart, wide-mouth mason jar
A weight

Chop the cabbage as finely or coarsely as you like. Mix thoroughly with salt. In a deep bowl, punch and squeeze the cabbage with your fists to extract its juices. You are creating brine. When you are confident you've gotten enough juice out to create enough brine to cover the cabbage in the jar, stop punching and pack the cabbage tightly into the jar with hands or a tool. Weight or airlock, allow to ferment. Some people un-crock their kraut as early as 2 weeks. I like to leave mine at least 1 month. Old World instruction would argue that real kraut is not ready until it has fermented 6 months. You make your own rules.

Apple Butter

I used to make apple butter every fall at the farm, on the same day that we had our pig roast in celebration of the first frost. Since then, the smell of smoking pork and spiced apple butter has been intertwined in my mind.

4 lb. cooking apples
1 cup unfiltered apple cider vinegar
2 cups water
Evaporated cane juice crystals (see notes for quantity)
Salt to taste
2 tsp ground cinnamon
1 cinnamon stick
½ tsp ground cloves
½ tsp ground allspice
1 Tbsp lemon juice, fresh squeezed

Slice the apples without peeling them and cook them in the water and vinegar over medium–low heat until soft. Puree. Per 1 cup of puree, add ¼ cup evaporated cane juice crystals. Add the spices, salt and lemon. Cook the apple butter uncovered in a wide pot over medium–low heat at least 2 hours, until nice and thick, stirring occasionally to prevent sticking. Pour into sterile mason jars, cool and refrigerate, or process jars in a water bath canner for 15–20 minutes.

Bread & Butter Pickles

Good on the side. Or for your pork and pickle pie, page 152.

2½ lb. pickling cucumbers
1 lb. onion, sliced thinly
¼ cup kosher salt
1¼ cups vinegar
1 cup cider vinegar
2½ cups evaporated cane juice crystals
1 Tbsp mustard seeds
1 tsp crushed red pepper flakes
¾ tsp celery seed
1-inch cinnamon stick
6 allspice berries + pinch ground allspice
6 whole cloves + pinch ground cloves

Slice the cukes into quarter-inch slices and place in a large bowl. Add the onion and the salt. Stir to coat the veggies in the salt, then cover the entire bowl with a tea towel and dump a few inches of ice on top of the towel. Chill the whole deal in the refrigerator for 4 hours. Then chuck the ice and rinse the cukes and onions. Drain. Sterilize 5 pint-size mason jars and lids. In a large pot, boil the vinegar, sugars and spices. As you are waiting for the brine to boil, pack the jars with the cuke and onion mixture. Once the brine boils, you can ladle it over the cukes and onions, until the vegetables are within 1 inch of the jar's rim, and the brine is within ½ inch of the jar's rim. Place the lids on the jars, cool, and refrigerate.

CHAPTER 4

Charcuterie

Charcuterie is the preservation of meat via dehydration, salt, cure or smoke. Some preparations call for only one of these methods, and many require the combination of two or more to ensure a safe and tasty end result. Charcuterie today is finally receiving recognition in the mainstream for the artistry that it is. Creativity abounds in the professional arena as whole-animal cookery returns to the American culinary repertoire. The fundamentals of charcuterie practice arise from many rich cultural traditions, and were established out of necessity long before the advent of refrigeration, or in situations where cooking or long-term storage were difficult.

I love thinking about how humans came across many of our foods (for example, who first saw an egg emerge from a chicken's backside, and decided to taste it?), but charcuterie is perhaps the most fascinating of human-made food manipulations. The best characterization I have ever heard of the impressive flesh experiments that have given us charcuterie was uttered by a professor at a guest lecture: "Basically, people decided to get a bunch of meat, salt it, stuff it in an intestine, and then throw it over their horses and ride. That's blind faith."

Deciphering Cured Meats

There may be a thousand ways to understand charcuterie fundamentals, but in my mind it tends to organize thusly:

- fresh sausage
- pâtés, terrines and meat specialties
- whole-muscle cures
  - cooked
  - uncooked
- fermented sausages

I list them in this order because this is how I think one should go about learning them, for two reasons: your skills will build somewhat intuitively this way, and your use of the carcass, as you become more experienced, will allow for (and possibly even require) an ascending knowledge and use of charcuterie practice. That being said, only a small amount of charcuterie practice is especially advanced. Mostly, it's just mysterious. To unseasoned cooks, this might seem like the same thing, but it isn't.

Charcuterie depends on the chemical and physical properties of salt and smoke, and on chemical reactions between curing agents, proteins, enzymes and even microscopic bacteria to preserve meat, in the meantime altering flavor, color and texture in ways that we often cannot predict until we sample the finished product. Herein lies the mystery, the adventure and the cautionary tale. Me? I automatically love any practice wherein humans must submit to the wild, unnamed, unknown workings of nature.

A note on safety: While I believe our culture is overly obsessed with sanitation, to the point where we have made infection more likely by eliminating both good and bad bacteria from our food and living spaces, one should take the potential hazards posed by botulism, trichinae, listeria and other hazardous microorganisms in meat processing very seriously. That said, clean work surfaces are possible without the use of ridiculous and harmful chemicals, and common-sense practices go a long way in the kitchen.

Bacteria deserve a great deal more respect from our species. As such, I would like to sing them an ode, and clearly state that bacteria are, without a doubt, the most important organisms on our planet.

They are our oldest ancestors (the original life forms), and without them and the role they play in the vast food chain that humans cannot see without a microscope, metabolic processes that both plants and animals require for survival would be strictly impossible. Period. Without bacteria, we all die. And without a proper respect for the power of certain bacteria, we literally will die. Louis Pasteur said it best: "In the end, it is the microbes that will have the last word."

Meat carries bacteria. This is a fact of life. So do vegetables and fruits, because all food is a product of nature, as are microbes. In the commercial world, it is accepted that poultry carries the highest bacterial "load," followed by pork, and then beef and lamb. In the butcher shop, cooler shelves are stocked accordingly. Poultry products go on the bottom to prevent their dripping upon anything but the floor. Pork goes above that, and the red meats go on top. Pork never touches beef, and poultry never touches anything. Vegetables stay with vegetables. It's rather easy. Don't go to the farmers' market, buy a chicken and a head of lettuce, and stuff them in the same bag. And the greater the surface area, the higher the potential bacterial load. Bacteria are opportunistic and reproduce quickly. Ground meat product has a greater total surface area and is therefore a greater substrate for bacterial life.

Knowing that bacteria are ubiquitous, and knowing that they are alive, allows us to discover that, like all life, bacteria have optimum requirements for survival. For example, *Clostridium botulinum*, the bacterium that causes botulism, thrives in slightly acidic, relatively warm environments with no oxygen. So in an aerobic environment (meaning oxygen is present), such as the outside of a casing, even if all other factors favor the growth of botulism, it still will not colonize. Where we do need to worry about botulism is on the *inside* of the curing sausage, and so we use nitrates and nitrites to prevent its toxicity. We know enough about the living requirements of the small handful of harmful organisms that threaten meat preservation that we can control them successfully. An understanding of the organisms and the tools we have, and a respect for inherent natural processes of cooperation and competition, allow for good charcuterie.

To counteract potentially hazardous microorganisms, we use several mechanisms. Above all, we strive to favor the growth and persistence of beneficial bacteria, which are not only responsible for the fermentation and curing processes—but their health and abundance also outcompete the harmful bacteria. Fortunately, with the use of some trusted tools, we can create the type of environment that ensures the success of beneficial bacteria. Doubly fortunate is the fact that these beneficial bacteria can thrive in a greater diversity of conditions than can the hazardous organisms.

Our trusted allies include salt, temperature control, humidity control, smoke and nitrites/nitrates.

Salt

There could be a whole book on salt and what it does to food and our bodies. Salt is crucial to the preservation of meat because it draws water out of the flesh via osmosis. Bacteria require water for survival, so we talk about the level of *water activity*, expressed in units Aw, when curing meats. The water activity refers to the amount of water available in a product. So if a piece of meat is frozen, it has no water activity, because all the water in it is turned to ice. Once frozen meat thaws, the water that was in ice crystals melts and is freed, and the meat once again has water activity. In charcuterie, a water activity of 0.92 or less will be shelf stable. The lower the water activity, the longer the meat lasts.

Salt is a huge part of this process, as is time. Salt also has properties that limit or change the growth and metabolism of bacteria, further aiding in meat preservation. And of course, salt greatly affects flavor.

Temperature

Bacterial and parasitic populations generally thrive, or exhibit certain behaviors, within an ideal temperature range. By controlling the temperature (think heating with fire and cooling with a refrigerator as the simplest forms of this), we can slow or completely eliminate harmful microorganisms, and promote beneficial ones.

Humidity

As with temperature, all life performs at its fullest within an optimum range of humidity, or amount of moisture in the air. By controlling humidity in conjunction with temperature, we can further tweak the environment in favor of safety. Additionally, humidity is crucial to the even and proper drying of fermented products. If the humidity is too low, meat will dry too quickly on the outside, forming a hard ring and causing the inside of the product to spoil.

Smoke

Smoke imparts a distinctive flavor and some color to meats. In cases of hot smoking, temperatures can get high enough to cook the flesh. Smoke also contains compounds that will slow or halt the ability of bacteria to grow and interfere with the process of rot. As such, controlled smoking can dry and preserve meats even at low temperatures.

pH

pH is a measure of the acidity of a product. It is important in charcuterie mostly when you get into fermented sausages, because fermentation creates more acidic environments, and because bacteria responsible for curing and flavoring will do their work only within a certain pH range. The pH scale runs from 1 to 14, where 7 is neutral. Numbers decreasing from 7 toward 1 correspond to higher acidity. Numbers ascending from 7 to 14 denote increasing alkalinity (which is the same as lower acidity).

Nitrites and Nitrates

Sodium nitrite, often referred to as pink salt, TCM, prague powder, cure #1 or instacure #1, is commonly used in preparing bacon, corned beef, some fresh sausages and other cured products that require shorter curing times or are cooked at the end of the preservation process. Cure #1 is active at low temperatures and contains 6.25 percent sodium nitrite; the rest is table salt.

Sodium nitrate, often referred to as cure #2 or instacure #2, is used in the preparation of cured meat products such as salamis and capicola, which require longer curing times and are not cooked at any time during processing. Cure #2 likes slightly higher temperatures and contains 6.25 percent sodium nitrite and 4 percent sodium nitrate, the rest also being salt.

Nitrites help preserve meat by imparting flavor and color, slowing rancidity and preventing spores of *Clostridium botulinum* from becoming toxic. Nitrate (cure #2) breaks down into nitrite (cure #1), so both products eventually get us to the same place. The difference between the two is that nitrite reacts directly with the myoglobin proteins in meat and with bacterial proteins to preserve meat and prevent botulism. Nitrates also react with bacteria, but just to break down into nitrite. The purpose of cure #2, with its combination of sodium nitrite and nitrate, is to provide a slow release of nitrite into the equation. As present nitrites begin to react immediately, nitrates can be slowly breaking down, readying more nitrite for later and thus supporting a longer curing process.

There has been a lot of public attention on the use of nitrites in cured meats, particularly because nitrites in our bloodstream bind to hemoglobin and decrease oxygen in the blood. Nitrite can also bind with amino acids to form nitrosamines, which have received a lot of attention as suspected carcinogens. However, there are no studies specifically and directly linking nitrosamines derived from meat or food to cancer in humans. When weighing the debate, I find that many people are grossly uninformed. Nitrites and nitrates exist everywhere in nature, notably in vegetables, where they are particularly concentrated in stems. Interestingly, nitrate levels in leafy vegetables can increase depending on growing conditions; for example, lower light resources cause the plant to store more nitrate in its stems. It's safe to say that it would take more meat than celery to kill a person from nitrate poisoning, and well-informed sources demonstrate that one would have to eat close to 20 pounds of cured meat in one sitting to produce toxic effects on the body. This is because nitrate concentrations are so low in cured meat products, partly because of regulation and partly because it requires only a tiny amount to produce the desired preservation effect. In addition, nitrites are metabolized (broken down and changed) during the curing process, so they do not exist in high quantities in the finished product.

Additionally, if we are thinking of our bodies holistically, we cannot merely compartmentalize the effects of nitrites in our blood and run terrified from cured meat products. We are part of the nitrogen cycle, just like the plants storing and converting nitrates within their cells. Our bodies will tolerate the bonding of nitrite to hemoglobin in moderate doses. Consider also that, amazingly, the consumption of meat (and leafy greens, as it turns out) provides iron to the body. Iron is a major component of hemoglobin. What a wonder. I like to joke that a person seeking healthy, balanced living is allowed to take a few known toxins in and trust that the body is resilient and complex enough to deal with them. I like to drink whiskey; I like to wear mascara; and I use sodium nitrite.

One final note on nitrate is a mention of celery juice powder, touted in many natural cooking forums as a suitable substitute for curing salts. Celery juice powder is extracted from the celery plant, which contains high levels of nitrate. The urge to endorse a more natural source of nitrate is certainly understandable; however, celery juice extract is rather unstable and can produce inconsistent results in curing and coloring of meat products. In the commercial world, producers—not surprisingly—prefer stabler, more thoroughly tested curing salts in known concentrations. In my mind, it is fine to experiment, but I can easily keep the stakes in plain view: nitrosamines are unproven nuisances, but botulism is a known killer. Nitrates must be consumed in insanely large quantities to produce lethal effects, while death from botulism takes a miniscule dose of *Clostridium botulinum* spores. I advise people to go with curing salts and avoid celery juice powder.

The proper use for curing salts is as follows, no matter the recipe:

- Use 4 oz. of cure #1 per 100 lb. of ground meat. When working in smaller quantities, arithmetic that keeps this ratio intact will have you using 0.2 oz. of cure #1 (or about 1 teaspoon) per 5 lb. of meat.

- For whole-muscle curing, you can use four times as much cure per pound of meat, so up to 4 tsp per 5 lb. of meat.
- Federal regulation permits the use of 3.5 oz. of cure #2 per 100 lb. of meat. For 5 lb. quantities of meat, use 4 teaspoons.

Getting Started With Fresh Sausage

Traditionally, most sausage making involves a curing step. However, modern fresh sausages are not always made with nitrites, to an admirably delicious result. Pure traditionalists balk at such practices as amateur at best, but the beauty of whole-animal cookery is that you can do whatever you want. Fresh sausages will get you in the habit of establishing proper ratios of fat, meat and moisture, and familiarize you with most of the equipment you'll need to produce other types of charcuterie.

Start with a recipe that is roughly 70 percent lean meat to 30 percent fat. Based on that total weight, be sure to include 1.75 percent salt, 10 percent (or more) moisture, and whatever spices you plan to incorporate.

Lean meat: Use healthy muscle tissue. Always trim to omit glands, silverskin, muscle with bloodspots, cartilage, etc.

Fat: Use belly fat or the firm, white fat that runs along the back of the pig, the back fat. Do not use caul fat, suet, or runny and supersoft, subcutaneous fats.

Salt: Use kosher salt. There are different brands, and each brand will be slightly different per ounce or gram. If you are weighing all ingredients, and you are sure to stick to your ratios, your salt content should please even the pickiest salt snob. It is generally accepted that sea salt contains too many inconsistent variables and minerals to produce even cures and flavors; however, many die-hard charcuterie practitioners use it in their whole-muscle cures, in addition to curing salts (which include, by weight, mostly table salt). My advice is to start out with predictable, standard, kosher salt, and then experiment with other varieties as you gain experience.

Moisture: This can be water, but better if it isn't. Use wine, vinegars, liquors and liqueurs, stocks, cream, etc. Remember that the type of liquid you use, and the amount of dry herbs and seasonings you use with it, will affect your moisture needs. For example, the more dry spices you involve in the sausage, the more liquid you are likely to need.

In my experience, less-than-amazing sausage is usually caused by one or more of the following:

- Poor or uninteresting texture. You'll notice a standard grind and half re-grind approach for creating good texture and bind in these fresh-sausage recipes. You will also possibly adjust, based on your grinder. Powerful machines, dull or damaged grinder parts, or improper grinding temperatures can effect texture; you'll need to learn your machine and keep it in good working order to avoid messy composition.
- Too much salt. 1.75 percent is the amount suggested in my recipes, and is generally a winning amount for fresh sausages.
- Not enough moisture, resulting in a crumbly, dry eating experience. It isn't enough to follow the master ratios, you need to adjust based on the other ingredients in the sausage. The recipes here have been tested for moisture. If you do the arithmetic, you'll see that some have moisture right at the 10 percent mark but some have as much as 14 percent, to provide for the interplay between all the ingredients.

Preparation

Cut the lean and fat into strips or chunks about 2 inches wide and up to 3 inches long. Be sure to trim off all silverskin, cartilage, gland, blood clotting and any dry or damaged meat. Silverskin can bind up the grinder, and glands and soured meat can turn the sausage rancid.

I always season before grinding, mixing my dry spices and herbs in with the moisture to make a paste and then rubbing this onto the lean and fat trim, just like a marinade. Many people grind and then season, but I find that seasonings bind better, and texture is superior, if you season first. Then, if I can, I let the meat sit in the seasonings, chilled, overnight; however, this extra time is not necessary.

Grinding

Work with meat as cold as you can handle it. Open-freeze the seasoned trim on a baking sheet for an hour or so, and place your grinder's moving parts in the freezer as well. For best results, work with frozen meat, if your grinder will handle it.

Working with meat at room temperature can cause "smearing," which happens when the fat melts during grinding. This results in over-emulsified sausage, poor texture, and less-than-ideal food safety. Remember that ground meat has a greater surface area, so more breeding space for bacteria. Keeping it cold will keep it more sanitary, provide a cleaner grind, and ensure a good texture for the end product.

Once you've sent everything through the coarse grinder plate, send half of the ground meat back through the grinder, to get variation in the texture. More surface area is a good thing when it comes to flavor and proper "binding" of salt and seasonings to meat.

Always test the sausage by forming a 2- or 3-inch patty of the ground product and pan cooking it. Taste and adjust seasonings before stuffing.

Stuffing

For fresh sausages, natural hog or sheep casings (from the intestine) are the best bet. Hog casings are used for standard sausages, and lamb casings for breakfast links. There are of course other sizes (beef middles are larger and used for summer sausage and salami, for example), and there are synthetic casings of all sizes as well. I always use natural casings for fresh product. Natural casings come packed in salt, in units called "hanks" (100 yards). A hank generally stuffs about 100 pounds of fresh ground product.

Before stuffing, natural casings will need to be soaked overnight and then rinsed thoroughly to remove excess salt. A great trick is to fit the end of the casing over your sink faucet and run a gentle stream of water through its length. This will also allow you to see any imperfections or tears, which you can cut out or choose to work around.

Once you've rinsed your casings, get the stuffer ready and load all the sausage in. Crank the handle until the meat is just barely coming out of the horn (this will prevent you from stuffing a load of air into the casings at the very beginning of the process.) Now oil the horn with a neutral vegetable oil and fit the casings over the stuffer horn, bunching them up like stockings as you load them on. Keep them moist as you go. You may choose to keep a small bowl of water near the stuffer as you work, so that you can moisten the casings as they begin to dry out.

Start by tying a square or bubble knot in the end of the casing. Then crank the handle and begin stuffing, using your receiving hand to steady the casings as they fill, and to ensure even filling. You can prick out any air bubbles with a needle as you go along. Try not to keep the casings even with, or slightly above, the horn as they fill. Instead, keep the sausage angled down slightly. This will prevent the introduction of air into the sausage.

When you've stuffed all the meat, tie a bubble knot at the end to seal the sausage. To form links, mark off a 6- to 8-inch length of the stuffed sausage, pinch, and twist the casing. Go up another 6- to 8-inch pinch and twist the casing in the opposite direction. Alternate directions as you pinch off links. This will prevent the links from unraveling when you hang the sausage to dry or smoke.

Drying

Drying the sausage before cooking it or smoking it is important. Dry products cook better, and the drying time allows the flavors to bind thoroughly. Drying the sausage also forms a "pellicle," a sticky coating of proteins on the surface that allows the sausage to readily pick up smoke.

Cooking

You can smoke, oven roast, poach, pan-cook or grill your sausages. If you're not smoking them, the best method for retaining moisture is to poach the sausages in water just below boiling point (170–190°F), and then finish them on a grill or in a well-seasoned, preheated skillet.

Breakfast Sausages

40 oz./2½ lb. pork lean trim
16 oz./1 lb. pork fat trim
1 oz. kosher salt
1 oz. cane sugar
0.5 oz. rubbed sage
0.5 oz. fresh garlic, grated on a microplane grater
0.4 oz. tsp thyme
0.2 oz. black pepper
0.2 oz. red pepper
0.25 oz. dried rosemary
0.2 oz. fresh grated nutmeg
½ cup white wine or fruit juice such as pineapple

Rub all ingredients into the pork, then allow it to chill overnight. When ready to grind, open-freeze the marinated meat on a sheet pan in the freezer, about 1 hour. Place your grinder parts in the freezer to get them good and cold. Fit the meat grinder with the coarse grinding plate and grind all the meat. Then send half of the ground meat back through the grinder a second time.

Soak and rinse about 12 feet of hog casings, and prepare your stuffer. Stuff sausages and allow them to sit overnight in the fridge before smoking or cooking.

Chorizo

Ancho chilies are dried poblano peppers and are prized for their sweet-hot qualities. Anchos can be found at tiendas *and many supermarkets and natural food stores. You can dry your own by taking fresh poblanos from your garden and roasting them in a 300°F oven until they are dried, then removing the seeds and grinding the peppers in your spice grinder until they are reduced to powder. The chipotle and ancho will deliver a mild spice to this recipe, but if you want more heat, use hot paprika, and consider adding cayenne to the spice mix.*

40 oz./2½ lb. pork lean trim
16 oz./1 lb. pork fat trim
1 oz. kosher salt
1 oz. dried ancho chili peppers, ground
0.4 oz. chipotle powder
0.4 oz. paprika
0.4 oz. fresh oregano
0.7 oz. fresh garlic
0.2 oz. onion powder
0.07 oz. black pepper
0.07 oz. cumin
¼ cup pork stock
¼ cup apple cider vinegar

Rub all ingredients into the pork, then allow it to chill overnight. When ready to grind, open-freeze the marinated meat on a sheet pan in the freezer, about 1 hour. Place your grinder parts in the freezer to get them good and cold. Fit the meat grinder with the coarse grinding plate and grind all the meat. Then send ½ of the ground meat back through the grinder a second time.

Soak and rinse about 12 feet of hog casings, and prepare your stuffer. Stuff sausages and allow them to sit overnight in the fridge before smoking or cooking.

Herbes de Provence Sausages

40 oz./2½ lb. pork lean trim
16 oz./1 lb. pork fat trim
1 oz. kosher salt
0.3 oz. fresh rosemary
0.2 oz. fresh savory
0.3 oz. fresh thyme
0.3 oz. fresh marjoram
0.3 oz. lavender flowers
0.2 oz. fresh tarragon
0.2 oz. ground bay leaf
0.2 oz. white pepper
0.2 oz. fresh parsley, minced
0.2 oz. fresh basil, minced
1 oz. fresh garlic, minced
1½ cups white wine

In a small coffee grinder or spice grinder, combine rosemary, savory, thyme, marjoram, lavender, tarragon, and bay, and grind until uniform and blended. Combine with salt and pepper. With a sharp knife, mince parsley and basil, and then add to the ground herb mixture. Rub all ingredients into the pork, then allow it to chill overnight. When ready to grind, open-freeze the marinated meat on a sheet pan in the freezer, about 1 hour. Place your grinder parts in the freezer to get them good and cold. Fit the meat grinder with the coarse grinding plate and grind all the meat. Then send half of the ground meat back through the grinder a second time.

Soak and rinse about 20 feet of hog casings, and prepare your stuffer. Stuff sausages and allow them to sit overnight in the fridge before smoking or cooking.

Garlic Orange Bratwurst

40 oz./2½ lb. pork lean trim
16 oz./1 lb. pork fat trim
1 oz. kosher salt
1.1 oz. minced fresh garlic
Zest of 1½ oranges
0.4 oz. white pepper
0.3 oz. ground ginger
0.3 oz. ground nutmeg
1 oz. Dijon mustard
1 cold egg, beaten
½ cup very cold heavy cream
¼ cup fresh squeezed orange juice

Grind and mix spices and salt, and incorporate into pork. Allow it to chill overnight. When ready to grind, open-freeze the marinated meat on a sheet pan in the freezer, about 1 hour. Place your grinder parts in the freezer to get them good and cold. Fit the meat grinder with the coarse grinding plate and grind all the meat. Now, in a small bowl, combine the eggs, cream and orange juice and work them into the meat with your hands. Then send half of the ground meat back through the grinder a second time.

Soak and rinse about 20 feet of hog casings, and prepare your stuffer. Stuff sausages and allow to sit overnight in the fridge before smoking or cooking.

Pâtés, Terrines, and Meat Specialities

Take your forcemeat skills to the next level by trying your hand at pâtés, terrines and mortadella. These preparations are essentially sausage making, but the products are molded, are served cold, and have more forward seasonings and more refined textures. In general, water cooking is the norm with these meat specialties—either poaching or baking in a water bath or pressing and molding are an important part of presentation and eating experience.

It's one of my highest dreams for Americans to embrace pâtés as a central part of our eating. They are not only beautiful and versatile, but can be a very economical way to eat meat. Pâtés are also an easy way to consume organ meat, which is very nutritious. You can get very creative with pâtés and terrines, and fancy yourself very thrifty indeed, as they can be made with meat trim and offal, and then accented or garnished with beautiful fruits, nuts, vegetables and more. Lastly, they must be made in advance to allow for chilling and molding, so they are great for entertaining, or having delicious food ready when it is needed in the home.

The only thing you need to make pâtés is a meat grinder, plus a few loaf pans. You can get pretty fancy with terrine mold pans, but for starters there is no need for that. Use your sausage-making principles (work it cold!) to keep fat, meat and moisture emulsified properly and you're on your way.

A lot of French and culinary terminology is thrown around when talking about this work, so below is a tiny glossary to demystify this very unmysterious subject:

- Terrine: French for *earthenware* or *porcelain,* meaning the product was poached or baked in a stone or clay dish. *En terrine* means that the pan itself molds the pâté, rather than being molded by pastry dough, the traditional method.
- Campagne: French for *countryside.* Used to refer to pâtés that are more coarsely ground.
- Au gratin: a portion of the ingredients (but not all of them) have been seared.

- En croûte: *in crust*. This is the oldest form of pâté preparation, wherein forcemeat is molded by being wrapped in pastry dough before cooking.
- Mousseline: *lightened with cream or milk*, and usually egg white, producing a very silky texture that holds the meat together.
- Aspic: a gelatin-rich meat stock that is clarified and added to terrines and pâtés to fill space, add flavor and suspend ingredients for visual affect.
- Panade: a flavorful liquid or paste, using breadcrumbs or flour as the thickening agent. Used to bind and flavor pâtés.
- Emulsions: dispersions of matter into liquid, usually made of components that are not soluble. In meat, it's easiest to think of emulsions as dispersions of meat, fat and water. The finest emulsions are those in which you cannot discern the meat from the fat from the water on your tongue.

Preparation

The process for preparing pâté is not far from the sausage-making process, but texture differs and seasoning is more aggressive. It is harder for the palette to discern flavor in cold food, so products served cold usually have higher salt content and more seasoning. There are no hard and fast rules for this, but the recipes included here will give you a sense of ratios. You alone will be the judge; you can always test and adjust seasonings after grinding and emulsifying.

Season before grinding, and let the meat chill with the seasoning for a while (overnight is ideal), then open-freeze before grinding.

Grinding

When emulsifying, it is even more important to keep the meat and fat very cold, so the fat does not melt and separate. As a rule, I work with frozen product when I am grinding anything; this way I worry most about grind and texture and less about temperature as I go. When working with pâtés, you will usually be grinding through the fine grinder plate, and sometimes sending product through a couple of times. In very fine-textured pâtés, you can follow up with emulsification in a food processor, or use an immersion blender. The more the product is communated (reduced to particles), the stickier and more elastic it will be, which produces divine texture and strong proteins that will glue the pâté together. Mixing also heats everything up, so remember to keep the mixture cold. Mixing too much will cause the emulsification to break, and ruin the product.

To test the flavor, prepare a sample by spooning out some of your mixture, wrapping it tightly with plastic wrap, and poaching it in 170-degree water until it is cooked through. To truly test it, chill the sample. If you don't have time for that, keep in mind that you'll want the seasonings to be slightly overbearing when you taste the mixture warm. The flavor will mellow as the product binds and chills.

Molding

Once you've got the texture where you want it, pour the mixture into your pan. Many people oil the pan and line it with plastic wrap, leaving about 2–3 inches of wrap overhanging on all sides. This makes it easy to mold the terrine or pâté after baking, and allows for swift removal of the molded product from the pan for serving. However, plastic wrap is difficult to apply without wrinkles, and the wrinkles can show up on your pâté, especially if it is of fine texture.

Once you've poured the mix into the pan, enclose the plastic wrap over it and cover it with foil.

Interested in layering mixes or suspending vegetables, nuts or fruits in the pâté or terrine? The process is simple and intuitive. Pour the first layer in and allow it to set slightly. Then add your olives or nuts or what-have-you, and then pour in more forcemeat.

Cooking

Pâtés are cooked in a water bath, and specialty forcemeats like mortadella and galantine are poached. Poaching is water cooking below boiling point (170°F is ideal). Water bath cooking entails placing the item to be cooked, in its pan, within a larger pan, and then filling the larger pan with water, so that the item is submerged about halfway in the water bath. Then the pâté is cooked until the internal temperature reaches a safe level (150 for pork, 160 for poultry).

Cooling

Cooling is an essential part of the molding process. As soon as you remove the pâtés from the oven, take them out of the water bath and put an even weight over them. I often use a bottle of wine or a heavy wooden cutting board. Allow the pâtés to cool on the counter, weighted, until they reach room temperature. Then transfer them to the refrigerator and let them chill completely, weighted the entire time.

When the pâtés are chilled and you're ready to serve, remove the weight, and invert the pan to turn the pâtés out onto a serving platter or board. Remove the plastic and you're ready to slice or spread, and pair with endless condiments and garnishes.

Liver Pâté

If you're timid about using offal, learn to love pâté. I put just about everything into it: hearts, lungs, livers, even kidneys. The beauty of the pâté is that you can create flavor profiles as easily as with sausage, and creamy, sweetened panades can mask the strong flavor of organs, turning out delightful sandwich spreads or an appetizer that will make it easy for anyone to eat the whole animal. If you don't like pâté, I urge you to keep trying. There are plenty of options for influencing the texture and flavor of pâtés to suit your palette, and as you go, you will find your palette will also change, to suit many exquisite pâtés.

2 lb. chicken or duck liver (feel free to combine pork liver, poultry heart or other offal)
¼ cup sweet yellow onion, grated
2 Tbsp garlic
2 Tbsp kosher salt
1 tsp fresh ground black pepper
½ tsp quatre epices (see sidebar)
2 Tbsp flour
2 eggs
2 Tbsp bourbon
½ cup cold heavy cream

First, oil two, 3-cup loaf pans or terrine molds, and line them with plastic wrap. Make sure the plastic wrap extends over the sides of the mold so you can fold it over the top of the pâté for cooking. Place grinder parts in the freezer. Preheat the oven to 300°F.

When grinding organs, best to deal with them frozen. I freeze liver in ice cube trays or deli containers; then I can send them right through the meat grinder or slice them with a sharp knife without making a gloppy mess (or strong smell) before grinding. Combine the frozen (or mostly frozen) liver with the garlic, onion, salt and quatre epices and grind, using the finest grinder plate you own.

Once the liver is ground, you may decide to emulsify it even more. This is completely optional. Very traditional *pâté de campagne* will have fairly coarsely ground pork shoulder in it, as well as some more finely ground organ. This recipe is designed to be pretty well emulsified, to produce a fine, smooth texture in the finished pâté. If you're going for smooth, it will probably be difficult to send it back through the grinder without much of it sticking to the machine, so I usually accomplish any further emulsion with an immersion blender.

When you're satisfied with the consistency, mix up the panade by combining the flour, cream, egg and bourbon. Quickly and thoroughly mix this into the ground liver mixture, then pour everything into the terrine molds, distributing the mixture evenly between the two pans. Carefully fold the plastic wrap edges up over the pâtés, and then place aluminum foil securely over the tops of both pans.

Next, prepare a water bath. Use a casserole pan or other vessel that is at least 2.5 inches deep and large enough to fit the two terrine molds side by side. Place the pâtés into the casserole pan and add enough water around them so they are sitting in a water bath that comes halfway up the sides of the molds.

Bake the pâtés in the water bath until their internal temperature reaches 160°F. (150°F is acceptable if you are using pork only.) Remove pâtés from oven and water bath and place an even weight over top. (I use my end-grain maple butcher block to weight multiple pâtés side by side.) Cool, weighted, to room temperature, then transfer to the fridge and chill, weighted, overnight. Serve sliced, with pickle, or spread on a sandwich, such as pork banh mi sandwiches (p. 148).

Quatre Epices

Quatre Epices is French for "four spices." It's a blend you'll want to have in your kitchen at all times, ready for confits, rilletes and pâtés, whenever the spirit moves you. It is very easy to make. The recipe below yields one cup.

⅔ cup white peppercorns

3 whole nutmegs, crushed under a heavy skillet

A heaping Tbsp of whole cloves

3 tsp ground ginger

Combine all in a spice grinder and grind to a fine powder. Store in an airtight container in the spice cabinet, where it is dark.

Headcheese

The most straightforward terrine is headcheese, as you produce gelatin-rich aspic while cooking the pig's head.

1 pig's head and 1 pig trotter, brined overnight in 1.5 gallons water and 1 box (3 lb.) kosher salt
2 whole, large leeks (including green leaves), rough chopped
3 large carrots, rough chopped
3 celery ribs, rough chopped
1 whole corm garlic, smashed to loosen cloves
½ bunch fresh marjoram
½ bunch fresh thyme
4–5 sprigs rosemary
½ bunch fresh parsley
6 bay leaves
1 whole nutmeg
6–8 whole cloves
2 Tbsp whole peppercorns
1 Tbsp whole coriander
1 tsp whole juniper berry
1 tsp red pepper flakes
1 bottle dry white wine
2 Tbsp balsamic vinegar
Generous salt and pepper, to taste

Remove the head and trotter from the brine and discard brine. Rinse head and trotter thoroughly in cold water and place in the bottom of a large stockpot. Add chopped vegetables and garlic. Bundle the marjoram, thyme and rosemary with twine and add to the pot. In a square of cheesecloth, bundle and tie the nutmeg, cloves, coriander, peppercorn, juniper and red pepper flakes. Place this sachet in the pot, along with the bay leaves. Empty the bottle of wine into the pot, then add enough cold water to cover everything by 6 inches or so.

Bring the pot to a simmer, uncovered, and cook gently for 4–6 hours, skimming regularly for the first few hours. When the jawbone loosens and the meat is tender, remove the pot from the heat. Remove the head and trotter from the pot and allow to cool. Discard the veggies, herb bouquet and spice sachet, and strain the cooking liquid (reserving it for further cooking). Return the strained cooking liquid to the flame and reduce by about half.

When the head is cool enough to handle, pick it clean of fat and meat, including brains and tongue (if desired—it's worth it). Smash the fat in your fists and pull the meat apart as much as possible. Add all to a bowl, and mix in the chopped parsley, balsamic vinegar, salt and pepper. Press into a terrine. Now ladle about ½ cup of the reduced cooking liquid over the headcheese, enough to just cover the meat in the terrine. Weight the headcheese and cool completely, then transfer to the refrigerator, still weighted, to chill thoroughly. Serve sliced or cubed, with mustard, pickles or any other accompaniments you can think of.

Bologna

Traditional bologna (pronounced, and sometimes spelled, "baloney") is cold smoked, but here in Appalachia, people seem to like it hot smoked. I agree with them. I also add liquor to mine, which everyone thinks is weird, but I don't care.

15 feet natural beef middles
80 oz./5 lb. lean beef trim
5.3 oz./⅓ lb. beef fat
¼ cup rum or bourbon
0.2 oz. cure #1
1.7 oz. kosher salt
2 oz. brown sugar
0.2 oz. ground allspice
0.1 oz. ground cloves
0.2 oz. ground ginger
0.2 oz. ground cinnamon
0.5 oz. fresh ground black pepper

Cube the beef lean and fat and mix with pepper, allspice, ginger, clove and cinnamon. Open-freeze, along with your grinder parts and a big metal bowl. Grind through the fine plate of your meat grinder into the frozen metal bowl. Mix the ground meat and fat with the cure, brown sugar, salt and rum, then grind everything again, this time through the coarse plate. (If you have a second frozen bowl to grind into, so much the better.) Cover and refrigerate the mixture overnight. The next day, prep your beef middles by rinsing them and soaking in lukewarm water for at least 30 minutes. Stuff the middles with the bologna mixture, as tight as you can, and tie links with twine. If you're a rock star, dry the stuffed sausages overnight again, and then smoke at 180°F until they reach 155°F. Plunge the sausages into an ice bath as soon as they get out of the smoker, then dry in the fridge one more day before serving.

Whole-Muscle Cures

Whole-muscle charcuterie is exactly what it sounds like. Instead of being ground or emulsified, the whole muscle is cured and then cooked, dried, smoked, etc. Examples are bacon, bresaola, prosciutto and guanciale. Whole muscles are preserved by dry curing, wet curing or a combination of both.

Dry curing is accomplished by rubbing the meat thoroughly with salt and nitrites, and often sugar or other sweeteners, herbs and spices. Once the mix is applied, the muscle is placed in the refrigerator and allowed to cure for a set number of days. The amount of time will depend on the size of the muscle and the ratios within the cure. Recipes usually specify, although a general rule is 1 day per pound for smaller cuts or boneless cuts, and 3–5 days per pound for larger cuts and bone-in cuts.

Wet curing is accomplished through brining. Some brining is done just to flavor and tenderize, but true wet curing is a preservation technique in which meat is brined in a solution of water, salt and nitrites. The meat is weighted under the surface of the brine and left to cure for a set amount of time. The timing will depend on the size of the muscle and the ratio of salt in the brine. Another method of wet curing is brine injection; there are various syringes and pumps available to accomplish this.

Wet curing and dry curing can happen together. For example, a ham may be spray-injected with brine, then dry cured and smoked.

Cure Recipes

MASTER DRY CURE
(per 5 pounds of meat)

3.5 oz. kosher salt
1.75 oz. brown sugar
0.2 oz. cure #1

MASTER BRINE FOR PORK, LAMB AND BEEF
(per gallon of water)

2 lb. kosher salt
Up to 1 lb. sugar or other sweetener (optional)
Spices (optional)
4.2 oz. cure #1

MASTER BRINE FOR POULTRY
(per gallon of water)

½ lb. kosher salt
¼ lb. sugar or other sweetener
Spices (optional)
4.2 oz. cure #1

Note that brine ratios are not set in stone. In many senses, the notion of a "master brine" is too simple: the more salt, the quicker the cure, and fish, poultry and red meat like different salt ratios. Professionals tweak their brines according to their intentions, and use brine tables and salometers, so they can set the *degree* or salinity of the brine based on the size of the meat and the type of muscle.

Also note that amount of curing salt differs in wet-cured products because it is calculated based on percentage "pick-up," that is, the amount of brine actually absorbed by the meat. Pick-up is estimated at 10 percent for the above recipes, but could be guaranteed only by injecting the brine into the meat, and measured for accuracy by weighing the meat after brining. When floating the meat in the brine (weighted until it is completely submerged, of course), as is recommended for starters, you can almost be certain that you will pick up less brine (and cure) than these calculations design.

In some cases, brining is used to flavor and tenderize (such as in the porchetta and headcheese recipes in this book) as opposed to curing the meat. In these cases, brining times are shorter and curing salts are omitted, since the muscles will be cooked afterward. To wet-cure items for cold smoking or fermenting, brining times are longer, and curing salts are added.

Bacon

This is a simple and delicious recipe for bacon. Dress it up with herbs and spices if you like. It is not necessary to remove the skin from a pork belly before smoking. The skin will keep it moist as it smokes, and is easily peeled off after the bacon is cooked.

80 oz./5 lb. pork belly, skin on
3.5 oz. kosher salt
1.75 oz. brown sugar
0.2 oz. cure #1
3–4 cloves garlic, grated on a microplane grater

Rub the cure thoroughly into the belly. Place into a ziplock bag and refrigerate for 14 days, overhauling daily. After 14 days curing, rinse the belly of cure mix and let it dry on a rack in the refrigerator before smoking at 170°F, until the temperature reads 150°F.

Smoking Meats

The smoking of meats arose in climates that did not favor air curing. In these cultures, smoke became an important step in curing, drying and flavoring meats. Smoking meats at home can be as simple as adding a firebox to your charcoal grill, or as elaborate as welding your own smoker together for multiple smoking techniques. There are three main ways to smoke, and each is used depending on your equipment, your intentions for cooking vs. exclusively drying, and many times, tradition.

Cold smoking is smoking with thin, intermittent smoke (the fire may go out once in a while), at temperatures no higher than 75°F. The fire itself is always kept at a distance from the meat product. This is called "indirect heat." Cold smoking is a very traditional method of meat preservation, and essentially works to dry the meat as opposed to cooking it. Cold-smoked products are usually smoked for longer periods of time, sometimes weeks. Speck is an example of a type of cured meat that is traditionally cold smoked.

Warm smoking is achieved with intermediate smoke, usually indirect heat, and moderate smoking times. Temperatures are usually between 80°F and 140°F. You can warm smoke a whole duck, or use warm smoke to prepare bacon.

Hot smoking is smoking with thick, consistent smoke, at temperatures between 145°F and 200°F. The aim is to cook the product while flavoring with smoke. Some practitioners will take temperatures much higher during hot smoking, and some even use direct heat. In general, though, low and slow is the idea when smoking, even when you are technically hot smoking. Sausages and tasso are examples of items you can hot smoke.

If you don't have any equipment, you can do something as simple as building a fire and devising a way to hang your meats above it. Many grills can be converted to smokers, and there are many ways to build smokers out of steel barrels, old refrigerators or even cardboard boxes. Some references are listed in the Resources section.

The wood you use for smoking is up to you, and you need not soak it before smoking, unless you are using an electric smoker. Different woods produce different smoke, so keep this in mind as you select your chips. Hickory is popular for thick, hot smoke, while apple is prized for fruity, moderate smoke, and beech wood produces light smoke. Other popular woods include cherry, pecan, alder, mesquite and oak.

Pancetta Stesa

Pancetta is a traditional Italian specialty that is cured and then hung without smoking. This is its chief difference from bacon. It is traditionally cooked before eating, but some folks eat it without cooking it first.

80 oz./5 lb. pork belly, skin off
4 garlic cloves, minced
0.2 oz. cure #1
2 oz. kosher salt
1 oz. brown sugar
1 oz. fresh ground black pepper
3 bay leaves, ground
0.2 oz. ground nutmeg
1 oz. ground juniper berries
0.4 oz. ground thyme

Mix the salt, sugar, cure #1, spices and herbs together in a small bowl. Pat the belly dry and then liberally rub it on all sides with the cure mixture. Place the meat in a nonreactive, closed vessel or bag in the fridge, and allow to cure for 9 days. Flip the meat in its cure daily (this is called "overhauling"). After the 9 days, when the belly is firm throughout, rinse it of the cure in cold water and pat it dry. Then wrap it carefully in unbleached cheesecloth and tie with butcher twine. Hang the pancetta for two weeks in a cool place in the house. You'll want it between 50°F and 60°F and at 60 percent humidity while it hangs.

NOTE: Pancetta may be cured rolled (arrotolata), or flat (stesa), as in this recipe.

Removing Pork Skin

Skinning pork is one of the best ways to practice knife skills. The easiest way to do it, however, is not always intuitive. Start, as you would imagine, by finding the most accessible corner, and undercutting the skin until you have a "tag" to grab on to. The object is to remove the skin while leaving as much fat on the meat as possible, and without leaving massive knife gouges in the beautiful layer of fatback.

Ready for the trick? Once you have the corner of the skin lifted, *flip the meat over* and press with your non-cutting hand, while your knife moves between the skin and the fatback. Keep the palm of your non-cutting hand flat on the top of the meat and keep the pressure even, while your knife keeps a steady, confident pull underneath. ◆

Prosciutto

Prosciutto is an air-dried ham that originated in Italy, where the climate especially favors its production.

1 whole ham (no sirloin) skin on, shank on, and bone in
Plenty of kosher salt
Ground black pepper
Rendered lard
Cheesecloth
A wooden crate or box to hold the joint, with holes or slats to allow breathability
A tray for catching moisture

Weigh the entire ham and record its weight. Then, inside of the box, completely bury the ham in kosher salt. Place your tray under the box. Allow it to cure in the salt 1 day per pound (based on the weight you recorded). Moisture may leach out of the box and into your tray as the ham cures. Keep an eye on the tray and dump it if you need to along the way. Be sure to keep the ham buried in salt, adding more salt as needed. When ready to hang the joint, rinse it thoroughly and pat it dry as well as you can.

Next, anywhere there is exposed meat or bone, thoroughly smear lard on the surface of the ham, then pack ground black pepper on top of the lard. Wrap the entire ham in cheesecloth, at least two layers thick, and tie it up in a cool, dark place. (I use the closet in my office, and many people use cellars.) Allow to hang for 6 to 12 months, or until it has lost 40 percent of its weight. When the wait is over, slice off very thin pieces and enjoy.

Capicola

80 oz./5 lb. pork lean, cut from shoulder, trimmed to 3 × 3-inch pieces
4 oz. kosher salt
0.2 oz. cure #2
1.5 oz. cane sugar
1 oz. ground black pepper
0.5 oz. ground coriander
0.4 oz. minced garlic
0.2 oz. grated nutmeg
0.1 oz. ground cinnamon
0.2 oz. lemon zest
0.2 oz. ground juniper

Mix the dry cure and rub half of the mix into the pork shoulder thoroughly, reserving the other half in a nonreactive container in the fridge. Place in a ziplock bag and refrigerate. Allow to cure for 9 days, overhauling daily. After 9 days, rub the second half of the dry cure into the meat and cure in the refrigerator for another 9 days.

When it's finished curing, prepare a beef bung by soaking it overnight in tepid water. When ready to stuff, rinse the bung, then stuff the pieces of cured shoulder into it, taking care to stuff tightly and eliminate air bubbles. (You can prick out air pockets by quickly puncturing the bung with a sterile sausage pricker or needle.) Tie a knot at the end of the beef bung and attach cotton butcher's twine to, or place a mesh net around, the capicola. Hang it in your curing chamber at 50°F–60°F and 65–75 percent humidity for about 30 days, or until it has lost about 30–35 percent of its original weight.

Using a beef bung: From the large intestine of a cow, this nonedible, natural casing will stuff up to 6 pounds of meat. It is usually about 20 inches long and 4 inches in diameter at its largest spot. Like other natural casings, it will come packed in salt. It is great for stuffing larger salamis and whole muscles you'd like to ferment. One drawback is that it is not of uniform diameter. If uniformity is important, you can trim it by taking off the narrower, closed end. (Save it to stuff something smaller later.) Tie a bubble knot to create a new end, and stuff as usual.

Lardo

Lardo is cured back fat. It is traditionally served sliced very thin on crackers, bread or sandwiches, but can also be cubed and used in sausages and salamis for further adventures in charcuterie. Just be sure to avoid light during the curing process, as light turns fat rancid. This is true for all cured meats, but when you're curing straight fat, as in the case of lardo, you'll want to be vigilant.

40 oz./2½ lb. pork back fat, trimmed of all meat, and squared off
1.75 oz. kosher salt
1 oz. cane sugar
0.2 oz. cure #1
0.2 oz. rosemary
0.5 oz. ground black pepper
0.2 oz. ground juniper berries

Combine the cure ingredients and rub onto the fat. Place in a ziplock bag and cure in the fridge as long as you like (at least 1 week), overhauling regularly. When you're ready to hang the lardo, remove it from the cure, rinse, pat dry and wrap in cheesecloth. Tie and hang at 50°F–60°F and 60–70 percent humidity for 2–6 weeks, or longer if you desire.

Smoked Fiochetto Ham

I've often heard practitioners of so-called culatello butchery *(page 132 in the Pork chapter) refer to the fiochetto as the "by-product" cut, where culatello is the prize. I like to think of the fiochetto as a little prize of its own. It makes an excellent smoked ham, and its size is ideal for the home salumist.*

2 gallons water
4–6 bottles of dark, malty beer
64 oz./4 lb. kosher salt
16 oz. sorghum syrup
16 oz. dark brown sugar
0.6 oz. juniper berries
1 oz. whole black peppercorns
0.6 oz. fresh rosemary leaves
2 whole corms garlic, smashed to separate cloves
5 oz. cure #1
1 whole fiochetto (4–5 lb.), lightly trimmed of fat and tied (see p. 134 for butchery instructions)

Heat the water and dissolve the salt and brown sugar. Cool, then add remaining ingredients except for ham. Chill the brine completely, then add the ham, weighting it below the surface. Brine for 14 days, turning the meat daily and ensuring it is fully submerged.

The day before you plan to smoke, remove the ham from the brine and discard the brine. Rinse the ham thoroughly, pat it dry, then place it on a rack in the refrigerator, uncovered overnight.

The next day, prepare your smoker for 150°F–180°F. A couple of hours before smoking, set the ham out on a counter to knock off the chill and, if you choose, set a small box fan behind it to dry it nicely before smoking. Smoke with oak and hickory wood until the internal temperature reaches 145°F–150°F.

Fermented Sausages

Fermented sausages are often regarded as the most complicated products of charcuterie practice. Recall that ground meat has a high surface area, and thus is a fertile breeding ground for microorganisms (both helpful and pesky). The inside of a sausage is a complex place. For starters, it is devoid of oxygen (botulism's favorite) and is being intentionally kept at temperatures ideal for the breeding of bacteria. For this reason, making fermented sausages requires the practitioner to both promote beneficial organisms and exclude those that may cause harm. This necessitates the use of nitrites, without question, as well as higher levels of salt, from 2.5 percent to 3.5 percent, to reduce moisture and keep the salt-hating pathogens at bay, all without creating an overly salty eating experience.

Ideal temperature and humidity are crucial in fermented sausages, to promote proper flavoring, texture and preservation via drying. For the most part, appropriate temperature and humidity can be achieved by keeping meats as cold as possible while making the sausages, and then by letting the fermentation and curing work occur in a charcuterie chamber with a controlled climate. (See Building a Charcuterie Chamber, p. 188.)

The other major factor to consider is pH. This is because pH level is both a product of functioning beneficial bacteria populations and a requirement for their performance in flavoring and curing the sausage. Remember that drying is what ultimately preserves the meat, and this occurs by decreasing water activity, thanks to salt and time. At the same time, fermentation—the process by which bacteria consume sugars and produce acids or alcohols—also occurs. The general requirements for this action are the same in sausages as they are in sauerkraut. The work of fermenting bacteria lowers the pH of the meat (makes it more acidic). This excludes harmful bacteria, alters the flavor and sets the stage for different bacteria to work on curing (converting nitrate to nitrite, as well as other reactions) and further flavoring.

Lactobacillus and *Pediococcus* are the chief genuses of bacteria responsible for the initial fermentation, which makes way for everything else. Their food is sugar, and as such, fermented sausages use different sugars to produce different ratios of fermentation bacteria. Temperature is crucial in the speed at which these species do their work, so very simply speaking, proper use of sweeteners, and temperature control, are vital to controlling the speed and extent of the fermentation process. Ferment too fast (higher temps) and the sausage might not taste right, because rapid acidification causes more intense souring. If you don't provide enough sugar, or the right kind, it will take longer for bacteria to produce the right pH for the next group of organisms to take over in curing the meat.

On the other hand, the bacteria that cause curing are species of two other genuses: *Staphylococcus* (that's right—there are harmless staphs, too) and *Kocuria*. These bacteria perform their work slowly, working with nitrate, metabolizing it to nitrite, and producing other reactions that flavor and color the sausage. This work happens at lower pH levels and at low water activity, once sugars have been consumed and *Lactobacilli* and *Pediococci* bacteria start to die off. Given this, it naturally takes time for these curing organisms to wake up and start feeding and reproducing, and the environment inside the sausage has to favor their preferences, which can be generalized as a pH of 4.8–5.3, and Aw less than 0.95.

Traditional fermented sausages were truly less doctored than our modern products, mostly due to available resources, partly due to ideal climate and partly due to the aforementioned blind faith of our ancestors. In cultures where the climate favored air-drying, fermentation was long and slow. Little sweetener was used, and the process relied primarily on wild cultures. The only inoculation of the sausage with any kind of bacterial culture was via a process of "backslopping," in which portions of a previous batch were mixed with a new batch to ensure proper starter cultures. To say nothing of the potential hazards and inherent inconsistency in that practice, it took much longer for products to reach the proper pH and Aw without more manipulation. Many practitioners in Europe still use these methods, preserving Old World technique while producing some of the most superior flavors.

There's a lot I love about that old way, especially as a curator in our modern era of fast-produced, cheap and monotonous food. I like food that takes time and love, where each effort is different and you never quite know how it will end up. Maybe that's why I love charcuterie in general: in a mass-produced world, it is artisanal through and through. Even still, the fermented sausages that dominate our market today taste different, because they are manipulated to produce a quicker and more consistent end product. Fast fermentation is accomplished by introducing freeze-dried starter cultures, acidifying agents, specially formulated combinations of sugars and, in general, higher temperatures. Because flavor-forming bacteria work slowly, and enter at the end of fermentation, a faster fermented product simply will not have the flavor of a traditionally cured sausage that has favored a slow process. More aggressive seasonings and careful combinations of sweeteners are used to flavor modern fermented sausages as a result.

For the beginner, working with modern recipes is a good starting place. You can test your product without as long of a wait, and can guarantee safer, more consistent results. I think it's valuable to try to limit additives (with the exception of nitrate) as much as we can, especially when working on the home scale, but for debutants, it is good to embrace starter cultures and special sweeteners. That being said, it's important to understand how all the added ingredients effect fermentation and curing, so you can do it effectively, or at least understand what you're reading on the label of your purchased cured meats.

Starter cultures are freeze-dried bacteria strains that can be mixed with sausages to ensure proper inoculation. In this way, you're sure to get the fermenters in there, and in chosen combinations based on what you're doing. Next, higher temperatures produce faster fermentation, and fermentation is what gets the product to that ideal pH and Aw. Sometimes an acidifying agent, such as *citric acid* or *Gdl* (glucono-delta-lactone), is used instead of higher temperatures; it's also possible to use both. Thirdly, you'll need sugar to feed the fermentation bacteria. *Dextrose,* straight glucose, is the most commonly used. It is a very fine-grain, very sweet product, so it is easily dispersed and fast to ferment. It comes from corn, so note that many of the concerns over responsible seed and land management that apply to our animal feeding standards will also apply here. Dextrose is guaranteed to be from genetically modified corn, unless you are sure to order organic dextrose. Sometimes, curers will use additional or alternative sweeteners, such as sucrose, lactose (sometimes in the form of dry milk powder) or maltose. The choice depends on the effect desired. For example, lactose and maltose can be used to slow fermentation and retain their sweet flavor, if one is concerned about too sour a sausage from rapid acidification.

Other ingredients you might encounter include *erythorbic acid* and *ascorbic acid* (*vitamin* C), used to impart desirable color; *soy protein,* for moisture and texture; *MSG,* for flavor; and *potassium sorbate,* for limiting mold. Recipes here will not include these additives.

In general, when making fermented sausages, you'll want to add starter cultures, spices and any additives to the lean, ground meat first.

Then add the fat, and finish by adding the nitrate/nitrite and salt. Mixing cultures directly with salt can produce inconsistencies in their performance, and getting in the habit of adding nitrate/nitrite at the end will help you avoid crazy chemical reactions between nitrite and some acid additives, should you choose to use them.

Note that humidity within the cabinet will be naturally higher at the beginning of the process, when the sausages have more moisture in them. Keep a close eye on your curing chamber when you first start using it. You may find that you need to manually tweak humidity between fermentation and drying, but you will also have to account for the changing conditions within the cabinet, and how they will regulate moisture levels naturally, at least somewhat. Also note that the larger the diameter of the sausage, the longer it takes to cure, so if you can make your first salamis in hog casings, you'll have a faster turnaround time, and less of a chance for humidity problems.

Basic Salami

Really, all one needs to make salami is meat, fat, salt, black pepper and nitrate. Maybe a little garlic. I say this to help you understand that you can actually create your own recipes, and nearly any fresh sausage can be morphed into a dried sausage, as long as you know the principles and proper ratios (2.5–3 percent salt, up to 1 percent dextrose, plus max. allowable nitrate). If you are using starter culture, you can choose what one to use based on how quickly you want the sausage to ferment; then choose how much and what type of sweetener you need. This basic recipe is very traditional. It calls for no starter culture and no special sweetener. It also has no fancy flavorings. For these reasons, I absolutely love it.

64 oz./4 lb. pork lean trim
16 oz./1 lb. pork back fat (or some lardo you've made yourself, for superior results)
2 oz. kosher salt
0.2 oz. cure #2
0.24 oz. cane sugar
0.5 oz. black pepper, or a mix of white and black pepper
0.5 oz. minced garlic
10–12 feet hog casings, rinsed and prepped

Open-freeze the fat and pork trim until it's just hard on the outside and still sliceable on the inside. Dice the fat or lardo into quarter-inch pieces. Grind frozen pork trim through coarse plate, into a metal bowl that's been in the freezer at least 30 minutes. Add frozen fat to meat and combine remaining ingredients. Mix thoroughly with clean hands. Stuff into casings and create 12- to 16-inch links. Weigh and record. Prick out any air bubbles, then hang to ferment in your curing chamber, first at 65°F–70°F for 2½ days at 85 percent humidity. Then reduce temps to dry at 55°F–60°F and 65–75 percent humidity until the sausage has lost 30–40 percent of its weight. This should take about 3 weeks. If you want mold on the outside, you can spray the sausages with a mold culture. Bactoferm™ M-EK-4 is a *Penicillium* strain that produces white mold. You can also cold smoke the sausage overnight between fermentation and drying. If you do this, don't bother with the mold culture, as smoke will inhibit its efficacy anyway.

Fennel Salami with Nutmeg and Wine

64 oz./4 lb. pork lean trim
16 oz./1 lb. pork back fat
2.5 oz. kosher salt
0.2 oz. cure #2
1.6 oz. dextrose
1.6 oz. cane sugar
0.5 oz. black pepper
0.25 oz. dried fennel seed
0.2 oz. minced garlic
0.2 oz. ground nutmeg
¼ cup red wine
0.1 oz. T-SPX starter culture

Freeze the lean and the fat separately, then grind the lean through the fine die into a metal bowl you've kept in the freezer for at least 30 minutes. Mix all ingredients except for salt and nitrate with the ground lean meat and refrigerate. Next, grind the fat through the coarse die into a separate, cold metal bowl. Mix fat into lean sausage mixture, then add salt and nitrate and mix thoroughly. Stuff into beef middles, then weigh and record. Prick any air bubbles and hang to ferment for 3 days at 65°F–70°F and 85 percent humidity. Then dry at 55°F–60°F and 65–75 percent humidity for about a month, or until the salami has lost 30 percent of its weight.

Pepperoni

40 oz./2½ lb. lean pork trim
40 oz./2½ lb. lean beef trim
2.4 oz. kosher salt
0.2 oz. cure #2
0.25 oz. cane sugar
0.2 oz. dextrose
0.3 oz. ground black pepper
0.6 oz. paprika
0.1 oz. coarsely ground anise seed
0.5 oz. cayenne pepper
0.4 oz. ground allspice
0.1 oz. T-SPX culture

Freeze meat trim, then grind the pork through the fine die into a metal bowl you've frozen for at least 30 minutes. Mix starter culture and sugars with pork and refrigerate. Grind the beef through the fine die into a separate chilled bowl and mix in all other ingredients. Combine the two mixtures, then stuff into beef middles. Weigh and record. Prick out any air bubbles and hang to ferment for 3 days at 65°F–70°F and 85 percent humidity. Then dry at 55°F–60°F and 65–75 percent humidity for about 2 months, or until the salami has lost 30 percent of its weight.

Building A Charcuterie Chamber

Fermentation of quality whole-muscle and sausage products requires a controlled climate, and you can create one fairly easily with a few purchased, and many recycled, materials. Using a standard refrigerator and attaching an external thermostat, an external hygrometer (for measuring humidity), a humidifier and a fan, you can be curing great meats at home in no time. This system allows you to maintain a climate of your choosing, and the temperature and humidity level you set it at depends on your project. For starters, it's best to aim for 50°F–60°F and 60–70 percent humidity in the cabinet.

There are a couple of options for outfitting your cabinet. One is to buy an external thermostat that you can simply plug the refrigerator into, and an external hygrometer that you can plug a cool-mist humidifier into. Then all you have to do is set the humidifier into the fridge, set a fan and some racks in with it, and you're on your way. This approach is more expensive, as easy, plug-in external thermostats and hygrometers can run upwards of a hundred dollars each. The other, cheaper option is to purchase a digital thermostat and digital hygrometer online, and actually wire your refrigerator and humidifier into those units. Below you'll find information on how to do that.

The thermostat pictured is the Lerway STC-1000; the hygrometer is the WH-8040. Both of these units can be purchased on Amazon. I learned about them from my friend Gred. He had rigged an STC-1000 and a WH-8040 to create a sweet potato dryer for the farm, and helped me re-wire them for one of my curing cabinets.

Some basic electrical knowledge. All circuits include a hot wire (usually red or black), a neutral wire (usually white) and a ground wire (usually green). The hot wire carries the electrical signal from the power source to the device, and the neutral wire returns it to close the circuit. The ground wire simply absorbs loose electricity from the circuit so that you don't get shocked when you touch it. When you're doing wiring work, make sure to close circuits, matching hot to hot, neutral to neutral, and ground to ground. For this project, you'll want to find an old air conditioner or refrigerator cord (3-prong, similar voltage to refrigerator) and expose its wiring. You'll wire from it to the external thermostat and hygrometer, and then from the thermostat to the fridge's power cord and from the hygrometer into the humidifier's cord.

Wire as shown in the diagrams.

It's as easy as that. Note that if you are going to keep your curing chamber in a garage

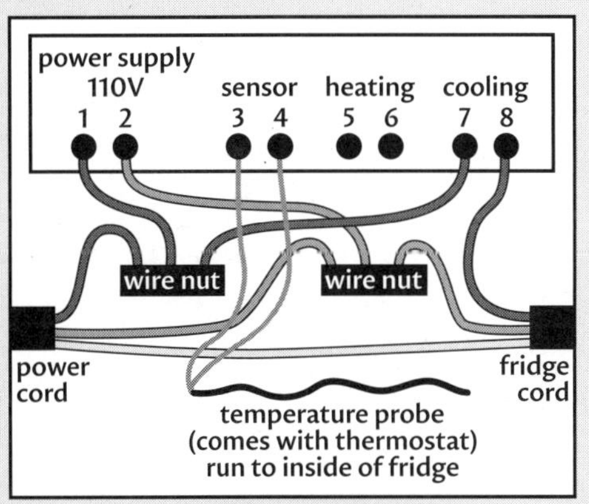

STC-1000

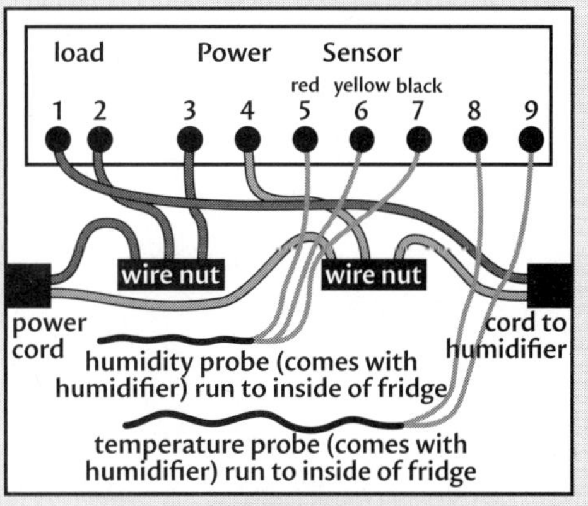

WT8040

or other non-insulated space, you may want to wire a heating pad into the unit. To do this, simply wire the pad into the "heat" nodes on the STC-1000.

Finally, you'll need to set the thermostat and hygrometer.

For STC-1000:

Press the S key and hold it down, then you'll be in set mode. The first setting is temperature in degrees Celsius, and you'll set it between 12°C and 18°C to start. To set, hold the S key and push the arrows to get the temp where you want it. Hit the Power key to save.

For WH8040:

Press the Set key and hold it down, then you'll be in Set mode. For the first setting, use the arrows to choose H for humidify. Hit Set again to save. Then hit the Up arrow twice, until you see LS on the screen. Press Set and then use arrows to put the low humidity setting at 55 or 60 percent. Press Set to save, hit Up until you see HS on the screen. Press Set and then use the arrows to set the highest humidity level at 65 or 70 percent. Press Set to save, and then Rst to exit Set mode. ◆

CHAPTER 5

Poultry

For as long as I can remember, I have had a hard time loving chickens, or any bird for that matter. I remember raising them on the farm: fat, white birds with expressionless faces and shit the color of the Ganges. Every bag of feed in the bin was like dumping money into a metal trap. Their overburdened crops, their slow, useless legs, their smell and their dust. I did not like raising chickens. And even though I was the fastest at evisceration on processing day, I silently dreaded it, and didn't even enjoy the thought of eating chicken at all.

Over the years I've spent in this industry, people struggle to understand why this is a problem for me. *You're going to kill it anyway and eat it. That's probably why you won't let yourself feel anything.* But they're wrong. On the contrary, I try to find the hard and salient living connection between everything, and connect it to myself, in pursuit of awareness. I don't think I ever found perspective with those birds, no matter how hard I tried.

I didn't fully realize this until I was sitting down to write about poultry, and I couldn't think of a single resonant emotion, other than disconnect. I felt terrible, and thankless, for all the chicken I've consumed in my life without having found much respect for the souls of these animals. This is not my way, and it surely isn't ethical. In the pursuit of the ethos of this book, what can I possibly say to excuse myself?

I have pondered and pored over this question, until I was forced to admit that I haven't got a clue. The only thing I know is that I never raised a true chicken. I raised Cornish Rock Cross hybrids, the darling of the cheap meat industry. I am appalled at the existence, and simultaneous weakness and invincibility, of this fat, white, industrial bird. It can hardly stand on its own legs, yet it props up a massive labyrinth of anti-food. And I cannot love it as a result. I can no longer eat it, either.

Yesterday, I visited my friends Brant and Shelly Bullock at the King Family Farm in east Tennessee, to ponder out loud about what can be done. We stood in the sunshine, flanked on all sides by heritage chickens, ducks and turkeys. We talked about the efforts of small farmers toward ethical poultry, and whether it is possible to achieve this within our lifetimes. Brant has incredible knowledge on heritage breeds, and carefully charts the economies around the breeding, rearing, slaughter and sale of carefully selected heirloom birds. The outlook? "It's dismal," he says, almost to the sky itself, in a voice of such exasperation that it borders slightly on amusement. I can remember that frustration. I don't miss it one stitch. Brant goes on to explain the challenges and joys of his investment in Buckeyes, Barred Rocks, New Hampshires, Black Australorps and other standard breeds. Their dark green, brindled brown, bronze and cream feathers flash in and out of wheatgrass pasture as we talk. They're pretty and fierce, these birds. They cock their heads to the side, looking at me as I look at them. Their eyes ask questions I'm asking too.

It's a mess, what growers are up against. There's no real chicken in the market at all. Even the most enlightened eaters scoff at real chickens, like the ones scratching here in Brant and Shelly's pastures. We haven't got a clue what chicken is anymore. The Cornish Rock Cross grows so big so fast, and is so cheap to produce, that Brant and farmers like him are looking for the first foothold. On top of providing customers with health, food safety, environmental and animal welfare education and cooking advice, and hatching purebred chicks or other chicks that Brant has developed, there's a sixth-generation farm here to run. A farm that has to work. Brant goes on to talk about hatcheries selling false heritage breeds, and the effects he suspects the industrial genetic experimentation in chicken is having on our children's health. I asked him and Shelly a lot of questions, but the main one is, "Why do it? In the absence of real answers, how do you stay so committed to this work?"

Shelly answered, "Brant really loves the birds. And because it's the right thing to do."

I don't mean to say that Cornish Rock chickens have no soul. They absolutely are living, breathing creatures. But they have no life. Even when we, as growers, try to put them on pasture, we see that they are not made for that existence. It's not admirable that I raised chickens for years without finding my personal connection to the enterprise, or to the animal itself. But it's also not surprising. When I was farming, I had no time for people who tried to talk to me about raising commercial, slower-growing, heritage birds. The economics made no sense, and still, now, there are precious few markets that will support the price they require. And so, as I sit here, preparing to extol ethical poultry, I feel tired, and pressured, grumpy and cold.

I think Brant and Shelly Bullock, and growers like them, are some kind of superheroes. In this day and age, my experience as a human is often one of looking around in a massive crowd and trying to find just one connection. In a huge, drone-like, pulsing, breathing, body of white monotony, these two farmers are like shiny bronze feathers flitting quickly through the crowd. To catch sight of them in such an uninspiring tapestry is enough to make me momentarily lose my breath. Farming is hard enough. We cannot expect America's entire farmer population to do what these people are doing: trying to reverse poultry's genetic demise, avert massive environmental and health epidemic, and build local economy, one heritage egg at a time. I peck and

I scratch for something else to say, something tangible and encouraging, but all I can say is this: *Every eater has to do more.*

And so, I wake up this morning, determined to right what I haven't yet gotten right, even though I have worked with animals half my life. "I need to get some heritage chickens," I say. "Just a few."

My partner, after an awkward pause, says carefully, "but...you hate chickens, Babe."

"I know," I say. "It's not OK."

Raising Poultry

Poultry, and specifically chickens, are the gateway livestock. These days, even urbanites are looking to keep a few laying hens and learning to harvest birds for roasting. It's no surprise. The poultry industry is infamous for animal welfare shenanigans, and confinement operations are troubling to food citizens who are concerned about everything from antibiotic resistance to the deepness of the orange in their egg yolk, and for good reason. I believe we do a whole lot of harm in the way we produce animals conventionally in America, but I think poultry takes home the trophy for the harshest hit to ethical meat. This is not only in the production standards, which involve crowded poultry houses, practices such as de-beaking to offset problems created by humans, and overuse of antibiotics, but also because of the full-scale ruination of the chicken itself through breeding practices that have created an ecological imposter and that mock enjoyable poultry eating.

Poultry is our favorite meat, and chicken is the animal we consume more than any other. Per capita consumption of chicken is a staggering 80 pounds. America harvests 8.5 million birds every year, equating to more than 50 million pounds of meat. Of the entire poultry industry (including turkey, other fowl, and eggs), meat from chickens accounts for 70 percent of income to the industry. In other words, we love chicken. Data is measured in weight of boneless, skinless meat, so it is difficult to truly tell how much this love weighs (or wastes). What is certain is that we don't tend to eat the whole bird, but expect it in pieces, and this is costing us more money than necessary. It's also leading to an unhealthy expectation of each component part of a bird (how big can a breast get?), and contributing directly to the aforementioned genetic abomination that we call chicken.

Many people don't realize that there are birds raised exclusively for meat production and others reared exclusively to lay eggs. The favored bird for commercial production of chicken meat is the Cornish Rock hybrid, often called Cornish Cross. They are bred to put on weight insanely fast, thereby providing the market with birds that have eaten less, required less time moving and using resources, and taken up less space on the farm. This animal grows so fast that in as little as eight weeks, the farmer can be ready to slaughter and harvest a five- to seven-pound bird. The faster the chicken can grow, the less the industry spends on feed and other expenses and the less you pay at the store for boneless, skinless what-not. But all this comes at a price, including environmental and working hazards, sacrifices to animal welfare, and gloomy dietary implications.

The vast majority of poultry in our country is raised by vertically integrated industry, wherein one company owns the hatchery, the processor and the distribution network, and contracts with farmers to grow baby chicks into finished birds. The farmers agree to produce a set amount of birds for a company, and prescribe to use the production standards and infrastructure of that company. These contracts are appealing because they guarantee a sale, and come with perks like a company veterinarian and pre-rationed feeds and supplements, but stories of botched deals, debts on company-required farm upgrades, and other horrors circulate well in rural America. The farmer becomes a number, just like each of his chicks, and must be kept within strict parameters to ensure company profits. Meanwhile, the cards are stacked against him and his birds. The Cornish Cross is bred to eat and grow, and has little other instinct or capability. They cannot survive predator attacks. They cannot mate or breed. They often put on weight so fast that they develop leg deformities and are found staggering around in chicken houses, maimed. They are stocked at such densities (less than 1.5 square feet per bird) as to pose high risk for disease, fighting and smothering. The feed includes, for the most part, genetically modified corn, soybean and other grain, as well as animal by-products, minerals and filler.

What of nutrition? And what of taste? In our discussion of other meat, we've learned that vitamins and minerals come from the earth. Probiotics come from an animal's connection to the entire food web, and protein and good fat come from the animals living full, active lives. A Cornish chicken is detached from all of these conduits that feed healthy, vibrant food. You can taste it in the meat, which frankly, has little taste at all. Many food citizens at this point aren't even sure of the way chicken ought to taste, because the market is so bloated with huge, cheap meat. These days, we feel lucky to get a bird that has been fed responsibly, with organic grains and non-soy protein sources, but even certified organic birds haven't necessarily had the chance to roam on grass or open their wings to the sun occasionally. Good food is not just about how it affects our health, even though this is a paramount parameter. If the life of the animal involves constant anguish, we cannot claim ethical food, or even tasty food for that matter. And so we revolve, inside of a difficult problem, with the Cornish Cross at the center. This is the meat that is available, and this is what we are suddenly used to eating.

Standard breeds, meaning those that act like chickens should, take 12 to 20 weeks (or *more*) to mature and develop muscle. They are active animals, with stronger joints, heartier muscles and better blood flow. They flap, run, peck and scratch. They live, breed and take in their environs. They're worlds different from the fat, white birds that lie beside feeders and bloat like feathered balloons. Their meat is darker and firmer, full of flavor, nutrition and depth. But we are having a hard time bringing them back. The Label Rouge standard, of French origin, brings heritage breeds together with improved production practices as the most tangible current source of available ethical chicken, but even these products are misunderstood as purely gourmet, and thus out of reach. It's clear that even if farmers want to change breeds, and favor flocks that live and taste better, they are up against a massive economic blockade, and the innocent ignorance of consumers. Standard breeds, fed longer with fresh diverse feeds, allowed to roam freely and moved to access fresh pasture, bring dignity and quality back to poultry, but bring confusion to the current marketplace. Customers who have grown used to fat, blanched chicken meat at $2.99/lb. will balk at the more wiry, darker ancestor, the true chicken, at $7.00/lb.

We should not be so complacent, and we shouldn't move so predictably. This industry we support tastes no good, and cultivates processes that are directly hindering ecological, economic and social balance. What's to be done? This is a meatier problem than we face with any other livestock species. Chickens aren't only our most produced livestock, they are also our most botched. Production is insipid, feed is contaminated, their very DNA has been grossly rewired. More than ever before, we must become educated as to the importance of the true costs of our food, and take action, beginning in our kitchens and fanning out into our backyards, all while supporting measurable improvements to poultry production on our local farms. In a vastly complicated food web, one thing is abundantly clear: Chicken is not a problem that will be solved for us.

Happily, real chickens are easy and fun to grow on the homestead. They taste better, bring joy to adults and children, and provide benefits in the garden and overall ecology of the homestead (whether urban or rural). In an industrial chicken house, each bird's short lifespan becomes an environmental and economic liability, but on the homestead, poultry can become assets that contribute to nutrient cycling in the garden or yard. What a fortunate thing, to have such a worthy revolution sitting squarely in our court. Chickens are a small, beautifully feathered and accessible way for anyone to grab the reins and begin to take back real food.

Breeds

Our most measured approach, as we slowly tool this revolution, will be to raise chicken for ourselves, both for meat and eggs. In this effort, it's best to favor dual-purpose heritage breeds, which will deliver quality eggs and also mature and muscle well for the table. Barred Rock, Hampshire and Buckeye are some of these breeds, harvestable at 12 to 16 weeks. There are many others as well, like Dominique, Rhode Island Red, Black Australorp and more. The American Poultry Association lists all recognized breeds. Where you start might simply depend on your favorite feather color, but know that these old breeds have a lot of variation between them. In fact, the original Kentucky Fried Chicken was developed around standard breeds, and back in the day, people used to order their chickens from the butcher according to breed, based on what kind of fat content they wanted and how they wanted to cook it. (Wouldn't that be something?) If you are raising birds for meat production, perhaps start with the Barred Rock. If you'd like consistent eggs, look to Black Australorp. Regardless where you start, you should definitely consider sourcing quality chicks, true to the breed.

To find good chicks, you must find good local breeders like Brant Bullock, or strive to become one yourself. The Livestock Conservancy and the Sustainable Poultry Network are two valuable places to start. The Livestock Conservancy catalogs all of its member farms that breed genuine heritage animals, from ox to guinea fowl. The Sustainable Poultry Network is focused entirely on poultry, and also certifies flocks in many US states, according to the American Poultry Association's standards. Information on these organizations is listed in the Resources section, and you can use their online resources to find breeders in your area. It's best to start with a handful of certified chicks (perhaps 6 to 12 for a family of four people) and get comfortable with the breed, and the rearing of poultry in general. Later you can try incubating and hatching your own eggs, and growing your flock at home.

I don't mean to suggest that we haven't bred all of our livestock to grow and behave in ways that suit our standards of production, and for maximum profit. Cattle, sheep and pigs have all been selectively bred to produce meat with qualities that we favor. But the practice has gotten wildly out of hand with chicken and turkey. With beef, for example, we obsess over Angus, but this is not a breed that we created in a fortnight, and it can survive and breed on its own. It has even become posh to expect heritage bloodlines in our pork. When it comes to turkey, there has been a resurgence of heritage breeds, although they are seen as uppity, gourmet, for holiday markets, for times when people are ready to spend more for their meat. Ground turkey burgers and turkey bacon abound at other times of the year, and you won't find heritage turkey meat in those. The humble gobbler and the everyday chicken have a further league to conquer before we'll be seeing *Breast of Buckeye* on the menu. So as you endeavor to raise birds, be careful and intentional

Poultry Breeds for Commercial Production

Make no mistake: the backyard is the easiest and best possible place for us to adopt superior breed ethics and champion slower, more sustainable poultry production. For farmers, on the other hand, it's not so simple. We need to restore ethical production and favor hearty breeds without negating a farmer's ability to make a profit. The mighty few farms and chefs who have tried to replace Cornish Rock chicken with heirloom breeds, full stop, are hitting a hard learning curve in terms of the readiness of their customer, whose palette and pocketbook have grown accustomed to Cornish Cross. The biggest challenge in the small-scale, commercial farming community is how to educate the public and bring chicken gently back, without gouging farm profitability. The solution has been middle-of-the-road approaches to chicken farming, like Cornish Rock birds raised with more honorable standards, and Label Rouge products sold right beside Cornish Cross carcasses. This is working, to an extent, as customers are learning there is a difference, and farmers who sell direct to customers are gaining traction with breeds such as the Freedom Ranger, a bird that combines Cornish Rock genetics with older breeds, producing a chicken that matures in 9 to 11 weeks, with better survival and foraging tendencies. These are the best ways for small-scale growers to invest in change, but they may not be able to be as radical as some consumers, or Brant Bullock, would like, because these farmers are still trying to make a living in the current food economy. This is yet another example of how farmers seeking to raise ethical meat must continue to balance what the market will bear now with what we're striving for in the long run.

Full-scale realization of ethical poultry is everyone's responsibility. ◆

at the start. Breed selection is a delicate issue, and proper management at the backyard level can truly change the paradigm for the better.

Your homegrown chicken will look different, and it will definitely taste different. You must prepare it differently, too. All in all, the adventure may be alarming, as you begin to grasp how strange and foreign is the supermarket chicken compared to the lively birds you rear on your own. Be open, as always, and let your animals teach you. As Jim Adkins, my friend and leader of the Sustainable Poultry Network, says, "It took us seventy years to mess up chicken. It will take us a while to get it back." This is an investment, not only in the hard science of breed engineering, but also in our cultural understanding and culinary preference.

Housing and Fencing

Figure 5 square feet per bird in the laying quarters, and provide enough head room that animals can perch on cross-beams, as this is their natural tendency. Leaving 18 × 18 inches for each bird to roost will be plenty of space. They'll need enough height in the coop to account for roosting, if roosts are set at least 6 inches from the floor. Roosts should be made of wood, about 2 inches in diameter, and rounded for the birds' comfort. Nesting boxes should also be installed, enclosed to allow isolation, and filled with straw bedding. The birds will ultimately want to roam during the day, and their meat and eggs (and your land) will be better for it, so figuring space outside the coop is more about where you'd like them *not* to go. The aim is to keep them safe in the evenings, when their natural predators roam, and to prevent them from tearing up your gardens with their scratching.

Most people start with a house of some sort, where the chickens can return for shade, water, nesting and a good night's perch. Houses with wheels or skids that can be moved to fresh grass are ideal. From there, electrified poultry netting, or chicken wire with an additional, low-lying hot cable, will protect your birds from weasels, foxes, dogs and raccoons. If hawks are a problem, overhead netting may be required. Some folks bother with very little fencing indeed, allowing their birds to roam most freely, after training them to return to their safe house at night. Very simply, the more freedom chickens have, the better their meat and eggs. Their scratching tendencies and droppings are also beneficial to the gardener when used intentionally. Some chicken houses are designed to mimic the size of garden beds; when it's time to clear out a crop, the house is moved over the desired area.

By far the most popular design is the simple A-frame house, with a roof on one end where the laying happens (girls like privacy) and an open, wire-covered section for air and light. Wheels can be mounted to one end and handles to the other, making it very easy to wheelbarrow the house to another location. Perfect for home production, these designs favor a few birds per house. Larger designs, which can be pulled with mowers or tractors, can be found online, or devised by the crafty homesteader. I've listed a few references in the Resources section.

Babies require extra considerations, namely heat until they can feather out a bit. A heat lamp, suspended from above, will allow the wee ones to huddle together and receive adequate warming. As they grow, watch them to determine how to adjust the heat. If they start to shy away from the light, or spread out or pant, they're ready for you to raise the heat lamp slightly, or lose it all together.

To cover intricacies of housing other fowl would take more space than I have to spare, so just note that you'll want to read up about the natural tendencies of other birds you intend to rear. For example, quail need height in their coops and turkeys obviously require more square feet. If you're raising ducks or other waterfowl, you'll want to make sure they have access to a pool or other body of water, and be prepared to change their quarters out frequently, as they are messy. Guineas, ducks and turkeys all require a tad more protein in their feed, as well. Many reliable sources advise not to mix your species of fowl, but you can take advantage of the different feeding habits of ducks and chickens if you run them together, and many people keep chickens with geese and guinea hens. Running chickens with turkeys is not advisable because of a parasite that causes a disease called blackhead, which can be passed between the species.

Feed, Minerals and Water

Chickens eat about a quarter of a pound of food per day, and you'll want the following essential components in a poultry ration, in relatively equal proportion: green forage, protein/fat and carbohydrate. The easiest approach is to find a reliable local source of organic premixed poultry feed, with at least 16 percent protein for egg layers and 20 percent or more for meat birds. Still, bagged feed is not as good as fresh feed, and even if you choose pre-mixed bagged feed, you should give your chickens access to green forage and grubs to eat.

Another drawback to buying bagged organic feed is that it can get expensive, and must be used within two weeks. The black soldier fly grub is a promising source of protein and fat for chickens, and can be produced easily on the home scale, with a model much akin to worm composting. (See sidebar for details.) You can also sprout your own grains, as outlined in the pork chapter, for additional flavorful grain and forage supplement. Note, however, that when sprouting grains for poultry, you'll want to ration their intake, and include protein-rich legumes like mung beans, cowpeas or field peas, provided you can source organic seed. Grinding and mixing your own organic grains is a very worthy practice if you have the time and a good seed source. There are books in the Resources section to get you started in proper rationing.

If you are interested in feeding your chickens from your garden or mixing feed yourself, consider the following as dietary components:

Carbohydrates	Proteins/Fats	Green Forage
Cracked corn	Grubs and worms	Kitchen scraps
Barley, oats or millet	Sprouted seeds	Garden trimmings
Sorghum	Fish meal (use sparingly)	Alfalfa, clover, cowpea and other legume cover crops
Wheat	Cowpeas, fava beans, soybeans (roasted), or peas	Dandelion, sorrel and other wild greens

Chicks just hatched will need crumbled "starter" feed, usually about 15 percent protein. Best to start by delivering the feed on shallow pans or flat boards, so the chicks can access it easily. You can slowly acclimate them to the feeder of your choice as they get taller and begin to move about more freely. Newborn chicks should be gently scooped up and their beaks dipped in water, to give them a taste and train them to partake on their own.

Feed troughs and watering devices can be ordered from specialty suppliers, purchased at farm stores or fashioned from wood. Many farmers raising pastured birds build long wooden troughs, similar in shape to basic wooden toolboxes, with spinning bars above them to prevent the chickens from sitting and pooping in their feed. The bars double as handles, allowing the farmer to move the feeders from place to place.

Feed for home-raised birds can be put out in the morning and evening, when the animals are emerging from or returning to their safe house. Try to introduce wild, green and raw, homemade foods before anything else. Any scratch feed, which is whole-grain feed, should be given as a kind of treat, to stimulate scratching tendencies as well as the birds' digestion. Chickens will also require grit to store in their gizzard, which, if you are not pasturing them, you may need to provide in the form of granite stone. Birds swallow their food whole; after that it goes into the crop, where it is mixed with digestive enzymes and saliva; on into the stomach for further processing; and then into the gizzard, where grit helps break it down before it is passed into the intestine. The grit gets passed along too, however, so it must be replaced. Many premixed feeds have grit already in them.

Standard bred birds will want to forage for grubs in addition to the feed you provide, and you can get creative with how to provide for this tendency. A long time ago, I created an interesting double fence around my garden, sort of like a chicken racetrack. The fence closest to the garden kept the birds away from the crops, while another fence about three feet farther out created a three-foot wide run all the way around the quarter-acre plot. The idea was to allow the chickens to hunt for insect pests as they entered, exited and sought refuge at the edge of the field, and to keep the perimeter clear of weedy overgrowth. It worked decently well, although I imagine the birds ate their fair share of desirable insects as well as pesky ones. Another idea for incorporating feed strategies with space allotments is to grow crops specifically to attract pest insects. Then turn the chickens out on to these "trap crops" for a protein-rich snack, and a little tillage to boot. Chickens can be unleashed on compost too, and they will turn it for you as they scratch for critters.

If you don't plan to let your chickens roam, or even move them from time to time, be sure to bring green foodstuffs to them, including vegetable trimmings from your kitchen, clippings of lush cover crop from the garden, and extra leaves from harvested vegetables. These fresh foods contribute greatly to the nutritive content of eggs, which is even visible in the darkening of the egg yolk. Feeding your birds a diverse diet will also boost the quality and flavor of their meat.

Supplements such as oyster shells (for calcium and stronger eggshells) and flax meal (for omega fatty acids) are often mixed with feed, along with free choice minerals, which the animal requires for healthy growth. Always provide shade and clean, accessible water. Gravity-fed watering cans or chicken-smart drinking nipples can be ordered from specialty suppliers; some sources are listed in the Resources section.

Additional Considerations

Roosters: Many people wonder if they must have a rooster in order to produce eggs. The answer is no, lady chickens can do the job with what they've got, they'll just produce infertile eggs. If you want to hatch eggs, however, you will need a papa chicken, obviously: a rooster. Roosters can provide some measure of protection for the flock, and also keep hens from fighting too much. Most roosters are amicable members of the flock, if a tad loud. If you do by chance get a mean one, you'll have to enjoy him for supper.

Pests and Diseases: The best thing you can do to keep your poultry flock healthy is to keep it free of pests and diseases. Keeping the

Raising Black Soldier Fly Grubs as a Poultry Feed Source

Hermetia illucens, or black soldier fly (BSF), is an extremely promising feed source for livestock. The fly is not like the common housefly. It is not a vector for pathogenic bacteria because there is plenty of oxygen in soldier fly colonies (which also means that their colonies rarely smell bad, if managed properly). Soldier fly grubs also eat other fly larva, so they can actually help rid your property of pesky fly species. They don't bite, since adults have no mouths, and they don't sting. Soldier fly grubs (larva) are extremely efficient decomposers and can feed on really high-protein waste materials that you wouldn't normally think of composting. At Living Web Farms, where I work, I even put an entire pig skin in the soldier fly bin, and it was gone within the week.

The amazing thing about these flies is that their pupa (adolescent grubs) are 42 percent protein and 35 percent fat, which is perfect for fattening livestock, without relying on the traditional and controversial soybean. Better yet, the valuable pupae even remove themselves from the colony when they develop, so they basically harvest themselves for your feed bin. Furthermore, it is relatively simple to establish a colony on the homestead, because you can easily attract the adult females to lay their eggs on waste materials you provide. At Living Web, we fermented some field corn and spread it between layers of cardboard, and then set those into our BSF bin to attract mama flies. A short while later, we had eggs, and we are now running a commercial-size bin and a home-scale bin, both of BioPod™ design.

You can build your own bin using a bucket with adequate drainage and an auxillary side-bin for the pupae mass exodus. The BioPods™ are developed by a company called ProtaCulture to favor the BSF lifecycle. Once the colony is established, the flies can be fed cooked, spent kitchen scraps, straight manure, animal offal and more. They don't like leaves, grass or stalks, however—save those for your compost pile. The keys to remember are to not overfeed (this will take some trial and error) and to have adequate drainage, as the feeds BSF enjoy are on the moist side, but the animals can't thrive in an environment that's too wet. The BioPod ™ includes drainage infrastructure, to help remove excess moisture from the colony. Also make sure the colony is located in the shade.

For more information on getting started, see the Resources section. ◆

hen house clean and moving your birds often will limit infections. Proper temperature, air flow and light will keep your animals happy and healthy. In addition, be careful about biosecurity, which is the exchange of contaminants between your poultry flock and other flocks. Don't wear your work clothes and boots to another farm, and ask others not to bring their work boots into your chicken yard. Avian influenza is passed from flock to flock in this way, so staying vigilant about cross contamination is wise. To knock back parasite load in your birds, feed them a couple of times each year with diatomaceous earth, a supplement made from the bodies of crushed sea creatures called diatoms, which will slice through the soft bodies of parasite animals. Garlic and hot pepper in chicken feed will also drive out parasites but won't necessarily kill them, so if you use those methods be sure to move your birds often.

Broody Hens: Heritage or standard-breed chickens are more likely to have a strong mothering tendency, which means you may get broody hens. These hens want to hoard eggs and try to hatch them, regardless of whether they are fertile. They will sit constantly in one spot, stop eating and drinking, and get pretty sassy when you try to move them off of their nest. The best solution is to exclude them from the nesting area by putting them in a grassy run with food during the day and a safe (but still excluded) area at night, without bedding. Make sure broody hens are eating and drinking; when you see that their desire to sit has waned, you can re-introduce them to the flock.

Home Slaughter

Home slaughter may sound like a huge hassle, but it is really quite simple, very do-able, and educational to boot. Before harvesting your first bird for the table, you'll need to make sure you have the right setup. You'll need:

- **A sharp knife:** Keep a designated knife for slaughter. It must be very sharp, and it is best to keep knives for butchery separate from knives used for cleaning and killing. A well-maintained 5-inch stamped boning knife will work fine for poultry.

- **A five-gallon bucket with a lid:** For catching blood and offal. If you want to keep the feathers separate, you'll need a second bucket. You can bury the offal that you don't use afterward, or feed it to your soldierfly colony; if you absolutely must send it to the landfill, it's a good idea to line the bucket with a heavy-duty trash bag.
- **A turkey fryer, or large pot with a propane burner:** Fill this with water and begin heating it before you start, for scalding the bird. Scalding at 170°F–80°F for 10–12 seconds will make plucking the feathers a total breeze.
- **A polypropolene cutting board, and knives for evisceration:** Use knives other than the one you used to slaughter. Ideally, you can have a knife just for eviscerating (gutting), totally separate from slaughter knives and butchery knives. I have used a small, serrated, 4-inch stamped knife for evisceration, but I much prefer a buck or a boning knife.
- **A garden hose with good water pressure:** For rinsing the carcass occasionally.
- **Spray bleach solution and rags:** It's important to keep surfaces clean. Keep feces or innards away from work surfaces as much as possible, because you don't want what's inside the bird to rub all over its skin.

Withhold feed from the intended bird the night before. If you don't, the bird's crop will be full when you eviscerate, and this can make a real mess. It's best to exclude the bird(s) you intend to harvest from the rest of the flock for the entire prior evening, and definitely do the slaughtering far away from your flock's home base.

When your scalding water is ready and your setup is finished, begin by holding the bird firmly upside down. If you are working alone, you can cut a hole in the bottom of a five-gallon bucket, pull the bird's head and neck through the hole, and hang the bucket up. Pull down on the bird's head so the neck is exposed. The best technique for slaughter is to make two cuts, one on either side of the bird's neck. This severs the carotid and jugular arteries, without severing the trachea or the

Make a cut on either side of the bird's neck to open the carotid and jugular arteries.

Scalding to promote easy plucking of feathers.

spine; essentially, the bird loses blood flow to the brain immediately and is rendered unconscious, bleeding to death rather than suffocating (higher animal suffering) or enduring trauma to its spinal cord (tougher meat via a shock to the nervous system).

If you have two people, you can team up to hold the bird and do the cut.

Wait as the animal bleeds out. If you have made the proper killing cuts, there will not be an insanely wild fight. What you'll experience is nerve reactions, as the body lets go. When you see lots of struggle, it's probably a sign that you've accidentally cut into the respiratory system or the spinal cord. Once the body is still and there is no more blood emptying from the neck, it's time to scald.

To scald, submerge the body into the 170°F–80°F water for about 4–5 seconds, then pull it out. Submerge again for another 4–5 seconds, then pull up again. You can pull on some of the wing feathers to test how easily they free up. These are some of the toughest feathers to pull, so if they start to give, you can see how close you are to a good scald. You may need to submerge the bird a third time, for another 4–5 seconds, before you feel good about the scald. The key is to balance water temperature, time in dunking and number of dunks, so you don't damage the bird's delicate skin.

To pluck, simply hold the bird over your designated bucket and go at it. Don't be timid. Work quickly, and pull feathers by the handful, in the same direction that they're facing, and you'll achieve a good pluck in no time. Look over the carcass after you've removed the bulk of the feathers to ensure there are no remaining pinfeathers or broken feathers attached.

If you wish to remove the feet, do so now, by bending the leg joint and cutting right in between the joint. If you want to remove the head, do it next, removing it just below your original kill cuts. Now rinse the carcass thoroughly and ready your cutting board and knives. It's time to eviscerate.

Start by making a downward cut in the skin from the top of the neck toward the wishbone, so you can open up the neck and see the top of the breast. Your goal in this initial step is to separate the crop (a fleshy sack attached to one side of the breast) and the skin. Using your hands, gently pull the crop away from the skin and the breast muscle until it is hanging freely. When you've finished freeing the crop, use your finger to poke a hole into the carcass, starting at the neck. This will enable you to pull the crop and esophagus through with the rest of the innards, which you'll remove from the other side.

Now turn the chicken around, breast up, and make a hole big enough for your entire hand, between the tail and the bottom of the breast muscles.

Now reach inside the bird and begin to separate the innards from the body cavity. If you can, pull the heart out first. Save it for pâté. Now, simply scrape around the body cavity with your fingers, separating all tissue from the

Plucking.

Loosening the crop.

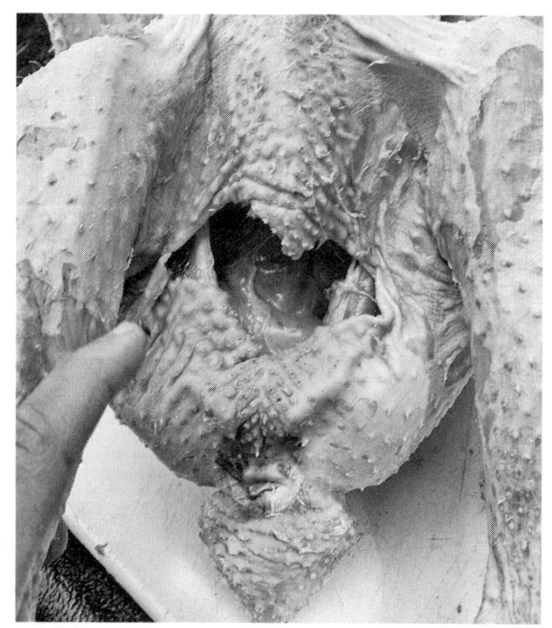
Preparing to dress the bird. Notice the torn skin on the chicken. This is from over-scalding.

body wall until you've gently gathered all the organs into a bit of a package. The esophagus will be the one thing that still feels attached. That's because it is running all the way up through the top of the carcass, and is attached to the crop.

Now gently pull all the organs out in a clump. You'll pull out the stomach, liver, intestines, gizzard and gall bladder, plus the organs nearby. Then you'll need to tug on the esophagus enough to pull the crop through the hole you made at the top. Pull hard enough to get it through, but try not to break it. If you do a little extra top work, you can sometimes pull out the trachea (a rough, ribbed tube running with the smoother esophagus) at the same time.

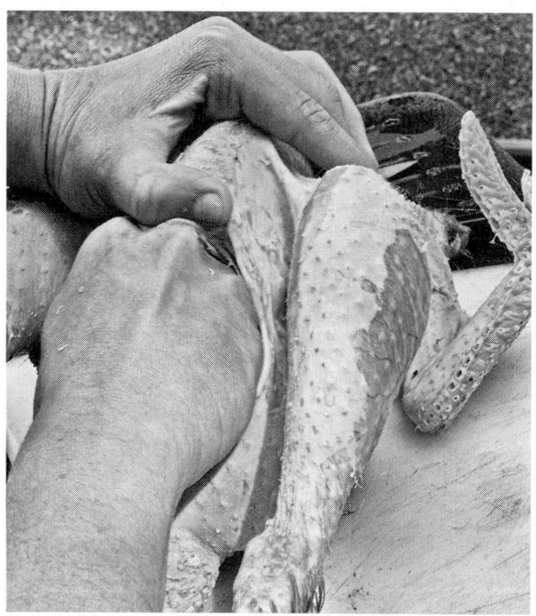
Use your hand to pull most of the organs away from the body cavity, into a package that you can remove all at once.

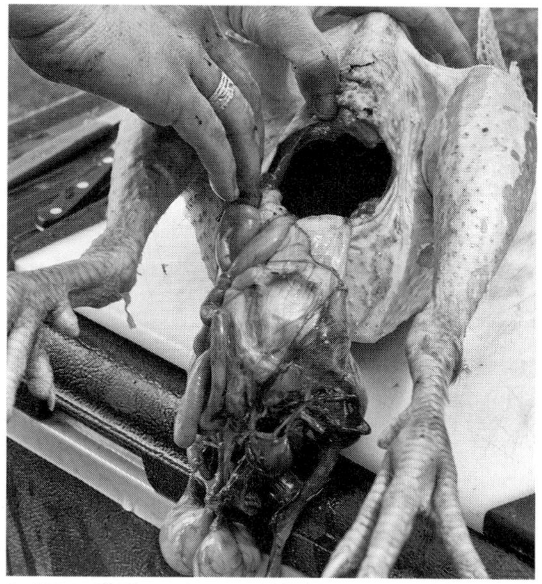
Pull out all the organs. In this photo, the darker organ is the liver (save for pâté). The crop is hanging below everything else, and you can see the esophagus attached.

To finish, you can either remove or cut around the tail, the oil gland that sits above it, and the organ mass all together, or you can make a cut to remove just the organs. You might choose to keep the tail intact if, for example, you want to truss the bird. To remove only the organs, make a cut around the cloaca (that's the anal opening), leaving the tail intact. This will ensure you've removed all fecal matter along with the intestines, and not left any hanging around inside or on the bird.

If you are leaving the tail intact, you'll need to scoop out the oil gland that sits above the tail on the outside of the carcass. Do this by undercutting the puffy area just above the tail and removing it completely.

Rinse the carcass thoroughly and refrigerate. Allow to chill at least 24 hours before cooking, so the muscles can relax postmortem. Otherwise the joints will be stiff and the meat tough.

Chicken Butchery

Chicken is a great place to hone your cutting skills. We'll cover the standard, eight-piece cut, and offer a slight variation, which will come in handy when you're frying chicken.

Begin by removing the wings. Stretch the wing as far as you can and move the joint up and down to find the socket. Using a semiflexible boning knife, cut the wing off at the shoulder, aiming for the space between the ball in the arm bone and the socket in the shoulder.

Variation/confession: I'm not big on the stand-alone chicken wing, so I like to cut the wing off with a bit of breast meat to accompany. To do this, simply angle your knife at about 45 degrees from the wishbone and cut off part of the breast as you remove the wing.

Next, free the oysters. This will assist you in pulling the legs off intact. The oysters are

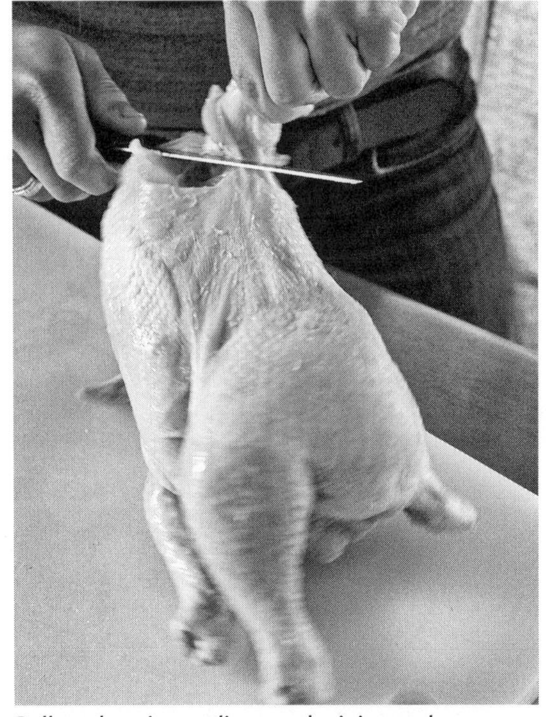

Pull on the wing to discover the joint, and remove.

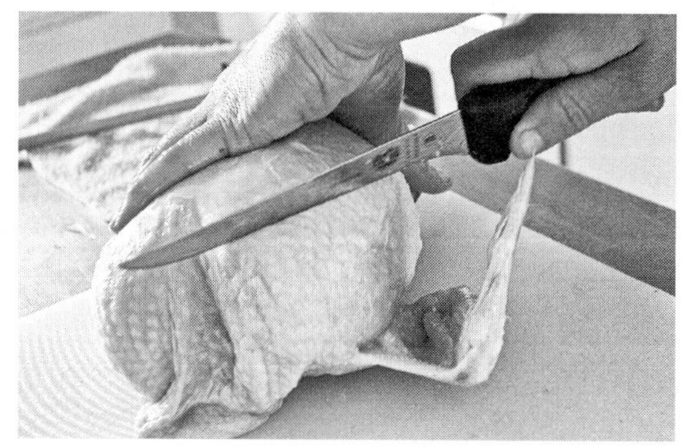

To remove some breast meat with the wing, cut at a 45-degree angle, downward from the wishbone, and then remove at the shoulder bone, as if you were removing the wing only.

Cut across the back to access the oysters.

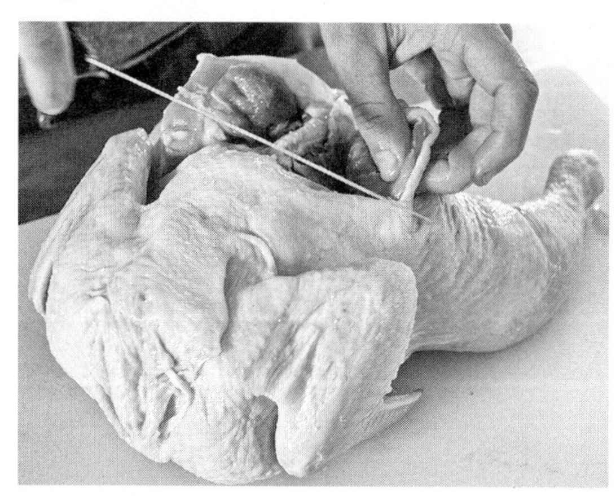

Freeing the oysters.

small, round, tender pieces of dark meat located on the chicken's back, one on either side of the spine. You'll find them right at the spot where the legs join the body, and this is where you'll make the first cut, across the back.

You'll see the oysters on either side of the spine. Use your boning knife to make small cuts close to the bone, undercutting the oysters and scooping them free. Leave them attached to the skin.

Now, flip the chicken back over so it is breast up, so you can remove the legs. Stretch each leg out at the hip and cut through the skin until you see meat.

Find the hip joint by gently wiggling the thigh to and fro, following accordingly with your knife. Pull back on the entire leg, peeling it away from the breast. You'll eventually pull the hip out of joint, after which you can follow through to remove the oysters along with the entire leg.

Next, divide the drumstick from the thigh.

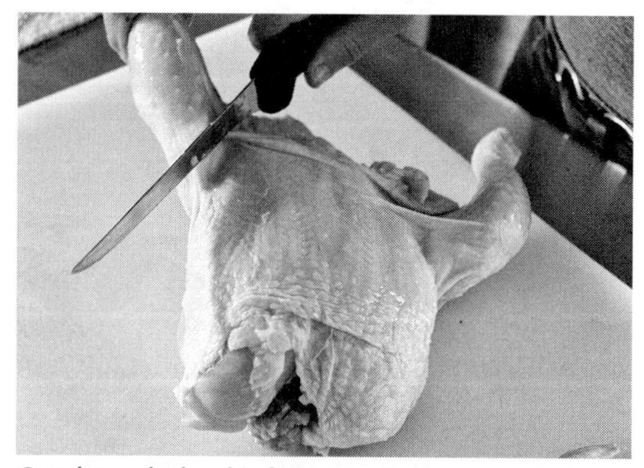

Cut through the skin between the breast muscle and the leg, to access the leg joint.

Pull back on the leg to dislocate the hip. Then you can easily complete the cut through the skin to remove the leg with the oyster attached.

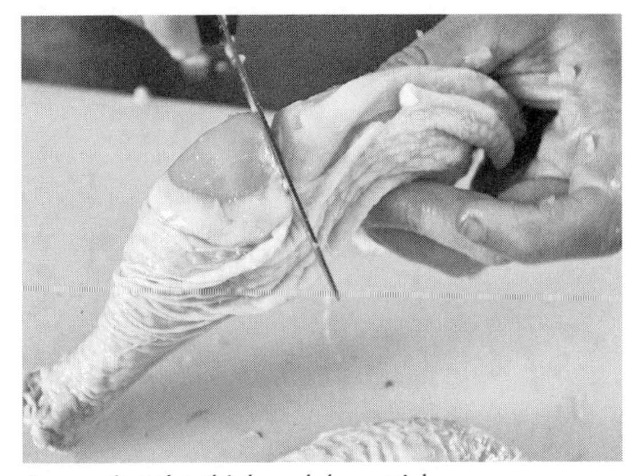

Separating the thigh and drumstick.

First identify the seam of fat at the joint, and begin your cut there. Once you've cut into the meat, you'll be able to see the joint clearly, and cut directly between the bones for a clean break.

The next step is to remove the backbone. You can use poultry shears for this if you're nervous, but it's not that tough to do with your boning knife. Just come at the spine at about a 45-degree angle, making close, downward cuts parallel to the spine. If your knife gets too vertical, you'll run into resistance, so as long as you keep that nice, slight angle, you'll be fine.

Finally, split the breasts. In between the two breast pieces is the sternum, or breast plate. I make straight cuts down either side of the keel in the breastbone and then start pulling the breast meat off of it with my hands, running fingers under the meat and close to the bone. Once you've exposed enough of the bone that you feel like you can pull it out, do it.

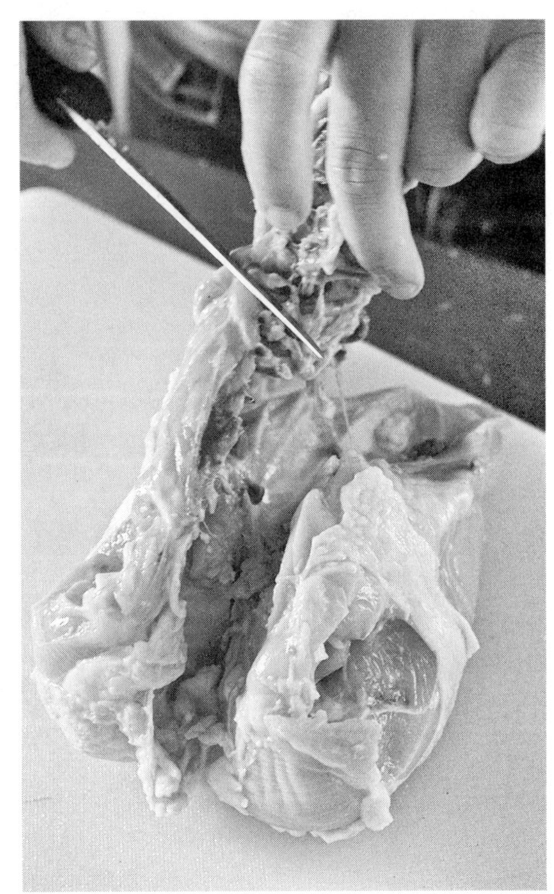

Keep your knife at an angle to remove the backbone.

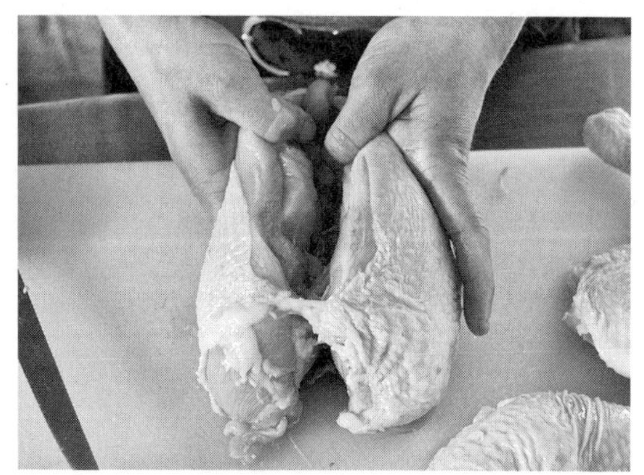
After making straight cuts along either side of the keel bone, begin pulling the breast muscles away from the bone, using your hands.

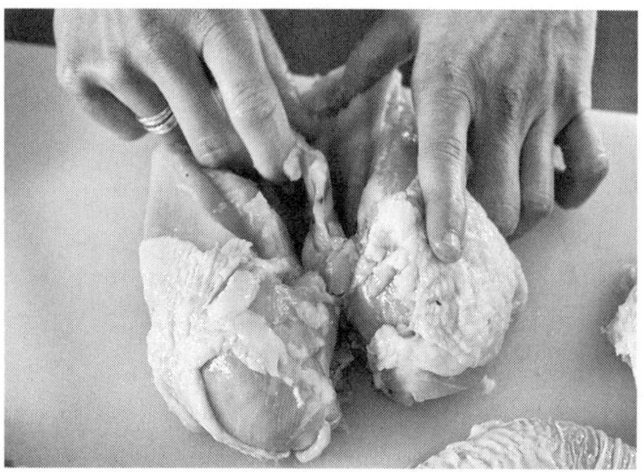
When the keel bone is exposed enough, pull it out.

Cut into the wing, between the bones, from top to bottom of arm.

You'll be left with breast meat that has ribs attached. You can remove the ribs by hand if you like, or use your boning knife to separate them from the back of the breast muscles.

Deboning

You'll use this skill for chicken ballotine, page 220—and honestly, it's the most efficient way to prep chicken for sausage. The first few times you do it, you might totally botch the meat, and that's fine for sausage making, because you'll be cubing up the meat and skin anyway.

Start by boning out the wings. Make a slit with your boning knife from the "armpit" to the "elbow."

Then break the bone at the shoulder joint, so that the upper arm bones are now poking out of the wing. Pull them out.

Next, remove the wing tip right at the joint.

Now it's time to address the core of the carcass. First, break through the center of the wishbone with your knife. Then make a straight cut along each side of the keel on the

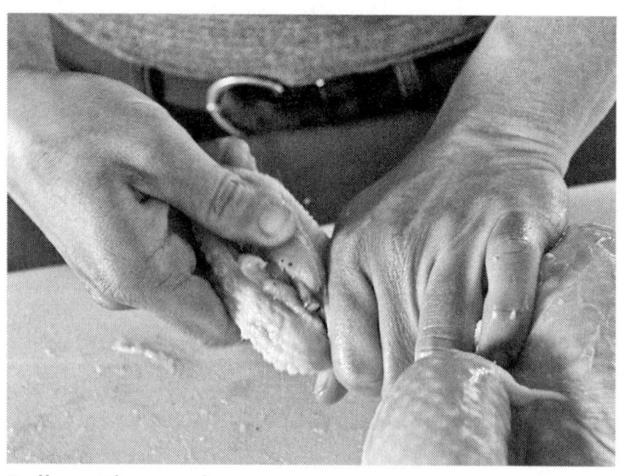
Pull out the arm bones.

Cut off the wing tip.

Straight cut on either side of the keel bone.

breastbone, from the wishbone to the bottom of the breast.

Then begin carefully removing the breasts and tenders with your hands, using small, confident strokes of your knife that ride close to the bone.

Remove the breast meat completely, and pull back on the shoulders until you can see the tendons and the shoulder joints. Slice through these with your knife.

Flip the chicken over. Starting at the neck, carefully begin peeling the skin from the carcass, using your knife to free the skin and your fingers to edge it off of the bone. Be sure you have properly severed the shoulder tendons and that the breast meat is coming off with the skin on the back, in one continuous piece.

If you're progressing successfully, the shoulder bones will be coming off with the rest of the meat, and all that will remain is the core carcass. Work down toward the tail until you reach the middle of the back. Then stop.

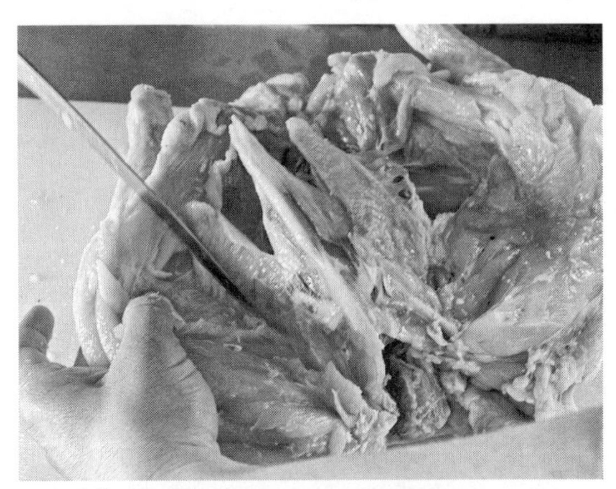

Carefully free the breast meat from the carcass, keeping everything attached to the skin.

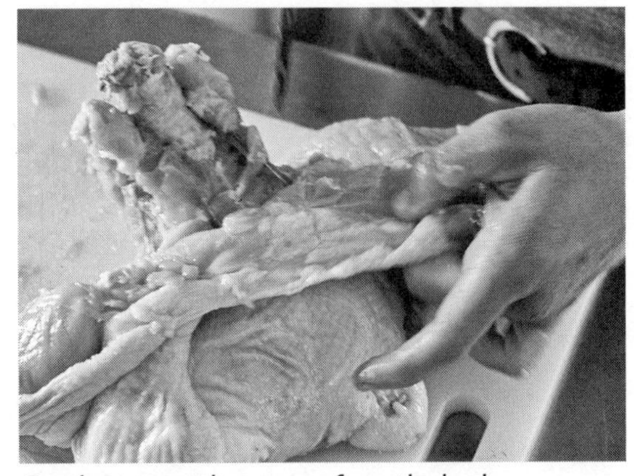

Gently remove the carcass from the back.

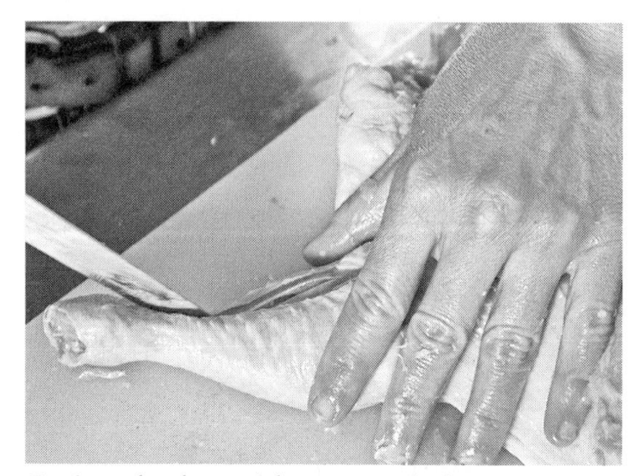
Cut into the drumstick to access the bones.

Removing the thigh bone.

Time to bone out the legs. Flip the chicken back over. Make a slit with your boning knife from the knee to the ankle joint.

Grab the ankle with one hand, and the rest of the carcass with the other, and bend the knee joint toward you to break it. The bone inside the drumstick should now be poking out so that you can remove it. As it comes out, cut away the scaly flesh right at the ankle, leaving it on the end of the bone.

Next, make a slit with your boning knife from the hip to the knee. Then, with your thumbs on top of the legs, and the rest of your finger underneath the thighs, bend the hip joint until you hear a crack, and the thighbone pokes out. Remove it, scraping downward toward the knee with your knife to relieve the bone of all meat.

Turn the chicken back over and continue very carefully severing the skin of the back from the carcass, until you have freed it completely. Be sure to save all your bones for stock!

Next, remove the cartilage from the knees,

Finish removing the core carcass, from the middle of the back to the tail.

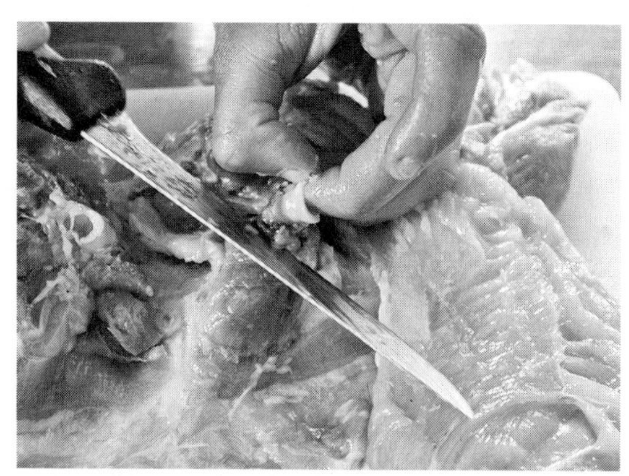
Remove the cartilage from the knee, and the pin bone from the leg with it.

and the tiny pin bones from each leg. You can do this in one punch by pinching the hard cartilage in each knee between your thumb and forefinger and undercutting it. As you come around to the place where knee connects to lower leg, you can undercut the pin bone that runs from knee to ankle, and remove it along with the cartilage.

Finally, you'll need to remove the wishbone, and the arm bones from the wings. To remove the wishbone pieces (remember, you broke it in half earlier), just feel around for them at the top of the breast muscles, then undercut them close to the bone.

To remove the shoulder bones, locate the top of each one and cut around its end, until you can begin to pull it up and away from the meat. You may have to scrape some meat from the bone with your knife, in a downward motion toward the wing, to clean the bone completely. Grab the end of the bone and pull. This action will probably turn the boneless wing inside out, which is fine. When you can't pull any more, cut around the other end of the bone to free it completely.

Check the meat for any remaining cartilage. If you've done an excellent job separating the skin from the back of the carcass, you're likely to have some remaining cartilage from just below the oysters. Pinch this and undercut to remove it, just as you did with the knee cartilage.

You're done! You've got a skin-on, completely boneless whole chicken. Time for gallantine, sausage or ballotine.

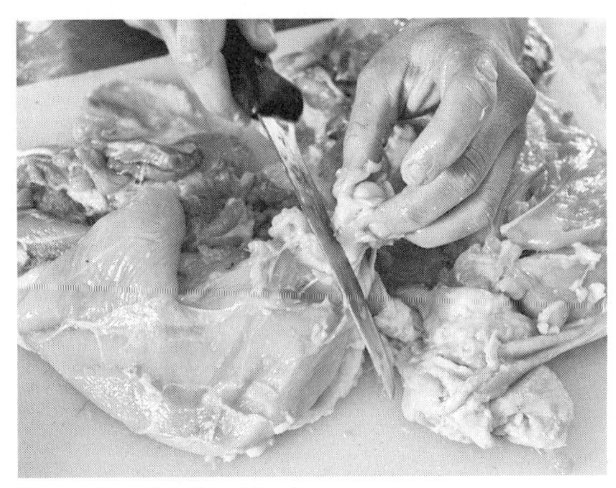

Removing the shoulder bones.

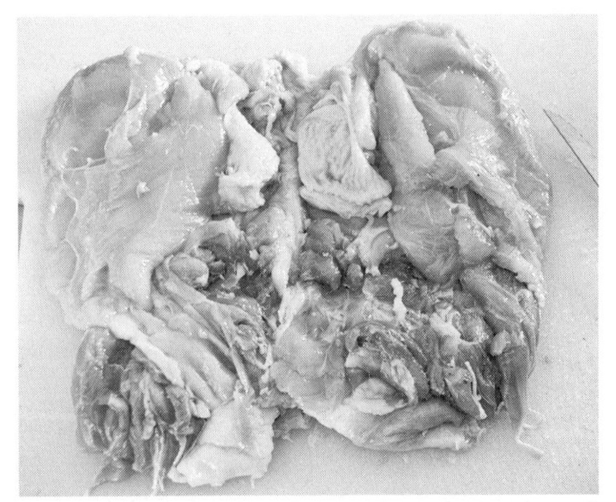

The finished, deboned chicken, from the inside. Pull the wings inside and arrange all the meat as evenly as possible.

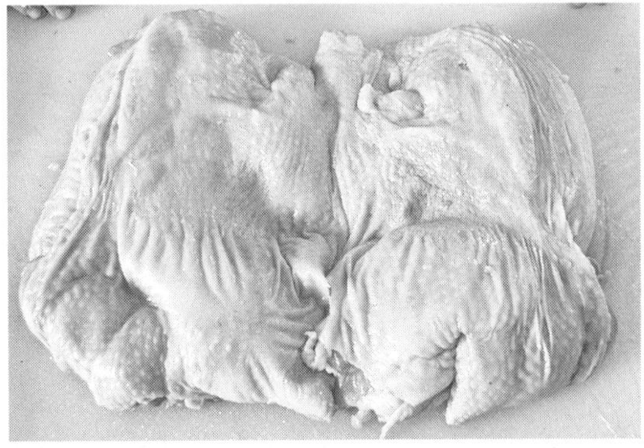

The finished, deboned chicken, from the outside.

Cooking with Poultry

Poultry is approachable. It is easy, it is light. We feel good about eating it, for rich or for poor. We eat it at the end of long, hard days, and we eat it on special days. It's comfortable. It's flexible. It's something we know. And believe it or not, we could probably know it better. My aim in talking about cooking is to encourage you to think about the whole bird. Cut it up as you will, in the privacy of your home, but leave no market, no yard or no store with chicken parts. It's the whole bird for the win.

Spatchcocked Roasted Chicken with Lemon and Basil

Simple, light and comforting, this roast chicken charms everyone. Choose a younger bird for open roasting. If you choose a matured animal, use a clay pot or enameled cast-iron Dutch oven with a lid, and cap it after you reduce the oven temperature. Cooking times may be longer for fully matured, heirloom birds.

1 whole chicken, backbone removed
2 whole lemons
2 cloves of garlic, minced
A dozen or so basil leaves, finely minced
¼ cup butter, softened
Sea salt
Fresh ground black pepper
½ cup dry white wine

Dry the chicken well with towels, then place, breast up, in a baking dish. Press firmly on the chicken's breast until you hear a slight crack, allowing you to lay the chicken flatly in the pan. Preheat the oven to 425°F.

In a small bowl, zest the lemons over the softened butter, then squeeze their juice into the bowl. Add the basil and garlic and mix thoroughly to form a compound butter. Next, carefully loosen the chicken's skin and rub the compound butter into the meat, in between the skin and the meat. You'll do this on the breast and on each leg, working to get the butter mixture evenly distributed. Sprinkle the skin with generous salt and pepper and place the bird in the oven. Roast for 15 minutes, allowing the skin to brown, then turn the heat down to 350°F and pour the wine into the bottom of the pan. Return to the oven and roast until a thermometer inserted in the breast reads 160°F. Allow the chicken to rest slightly before carving.

Fried Chicken

I've been studying fried chicken, and tasting a decent bit of it, in preparation for this book. My friend, and Southern food authority, Ronni Lundy, went with a fellow farmer, Walter Harrill, and me on a tour of places serving fried chicken on the bone. There were several approaches: brined and un-brined, deep fried or pan fried, battered with cornmeal and battered with flour. There was no real winner, but the standout approach was of the old-timey variety: an heirloom chicken, brined, dusted with dry batter and fried in a shallow pan at King Daddy's Chicken & Waffles in Asheville, NC. My recipe is similar, though not an exact replica, and calls for some extra cutting. Ronni explained that back in the day, chicken was cut into smaller pieces, so everyone could get more than one. I like this approach. Fried chicken is a treat, and it should be savored as such. Be sure to harvest young birds for frying; older animals will be too tough.

1 whole chicken, split with a bit of breast on each wing
1 quart cultured buttermilk
2 tsp kosher salt
1½ tsp paprika
2 tsp black pepper
1 cup all purpose flour, sifted
Sunflower oil for frying

Enameled cast-iron Dutch oven, with the lid

Separate the thighs from the drumsticks to further piece out your chicken. You can also divide the breast portions further, to create smaller pieces. Just take care to keep the precious skin intact, as its presence is integral to the perfect fried chicken experience.

Place all pieces in buttermilk and refrigerate overnight.

When it's time to fry, pat the chicken pieces as dry as you can. Combine salt, pepper and paprika in a small bowl and dust over the chicken pieces thoroughly. Allow to rest while you sift your flour and start the oil heating. You don't need deep oil, just give it about 1½ inches in the bottom of the pot. And the flour should for sure be sifted—even if you don't have an official sifter, just get your fine mesh sieve and knock the flour through it into a paper bag. When the oil is at 350°F, place a few seasoned chicken pieces at a time into the bag of flour, roll the top down and give it a good shaking. Remove floured pieces from the bag, turn the heat down from high to medium-high and place chicken pieces in the oil. Brown the pieces nicely on both sides, place the lid on the Dutch oven, and let her roll for about 20–25 minutes. If you've ever wondered how real fried chicken gets the perfect crisp, with the softer, velvety skin underneath, it's because it kept maximum moisture inside during frying. This is why you should put a lid on it and fry in shallow oil.

Remove finished pieces to a drying rack and set over a baking sheet to drain. Let the air hit them a minute, to dry them a bit. Serve with waffles and maple syrup, or gravy and biscuits, or mashed potatoes and mac and cheese...you know how it goes.

Chicken Ballotine, Three Ways

Ballotine is a boneless, stuffed, rolled and roasted whole chicken. It allows endless flavorful creations, and offers a beautiful presentation.

Prep for the chicken is the same each time. First, debone the whole bird, per the directions on page 214. Place the stuffing down the middle of the boneless carcass, and then tri-fold the rest of the carcass up and over the stuffing. Tie with butcher's twine every two inches, and salt and pepper the skin. Roast for 15 minutes in a 425°F oven, then pour a glass of white wine or chicken stock into the bottom of the roasting pan and reduce the heat to 350°F. Continue roasting until the internal temperature, measured with a meat thermometer, reaches 160°F. Slice ballotine into 1½-inch servings and serve hot.

Honey Mushroom Duxelles Stuffing

½ lb. fresh mushrooms, thin sliced
3 Tbsp extra virgin olive oil
2 cloves garlic, thin sliced
½ large sweet onion, thin sliced
3 Tbsp raw honey
1 tsp sea salt
½ tsp fresh ground black pepper
½ tsp red pepper flakes
juice of ½ lemon
2 Tbsp heavy cream
½ cup white wine or chicken stock

Heat the olive oil in a wide skillet over medium-high heat until it runs thin, then sauté the onion and garlic. Add mushrooms and salt and cook over medium heat until the mushrooms release their juices. Add black pepper, red pepper and lemon juice and stir to combine. Slowly add honey, stirring to coat mushrooms, and simmer lightly for 2 minutes. Add the cream in a slow stream, stirring to combine. Remove from heat and cool slightly, then spread evenly down the center of the whole, deboned chicken. Roast as directed above.

Olive, Feta and Sundried-Tomato Stuffing

12 oz. of pitted kalamata olives
1 whole corm garlic, cloves separated and peeled
1 cup extra virgin olive oil
6 oz. sundried tomatoes
8 oz. goat or cow feta
salt and pepper to taste
¼ cup toasted walnuts (optional)

In a small saucepan, lightly roast the garlic in the olive oil until light brown and soft. Drain the oil off, reserving it for later. In a food processor, place the sundried tomato, olives, garlic, salt and pepper, and a about ¼ the oil. Process to form a puree. If the mixture is not smooth enough, add more of the garlic oil. Salt and pepper to taste.

Place spread down center of a whole, deboned chicken, spread feta in a second layer, and walnuts on top. Roll chicken and tie according to directions above, and roast.

Cashew Cheese and Chili Pepper Stuffing with Pancetta

1 cup raw cashews, soaked for an hour or so in lukewarm water
¼ cup nutritional yeast
¼ cup chicken stock
Juice from one lemon
Splash of white vinegar
2–3 gloves garlic
1 Tbsp coarse ground mustard
1 Tbsp sea salt
2–3 fresh poblano peppers
8–10 thin slices of pancetta (or bacon)
Salt and pepper

While the cashews are soaking, place the poblanos on a baking sheet in a 375°F oven and roast until skins are blistered. Flip the peppers over and roast the other side. Remove from oven and cool on an open countertop. When they are cool enough to handle, pull on the stems to gently remove the mass of seeds. If you don't get it in one shot, gently rinse the peppers under a thin stream of water to remove all their seeds. Pat each pepper dry, open one side with your knife to flatten them, and sprinkle them inside and out with salt and pepper.

Drain the cashews and discard the soaking liquid. Put cashews, nutritional yeast, lemon juice, wine, broth, garlic and mustard into a food processor and process until creamy. If the mixture is too runny, add more nutritional yeast. If it's too thick, add more broth. Taste and adjust seasonings. I like to add pepper here.

Lay the roasted, salted peppers over the inside of a boned-out whole chicken, then disperse as much of the cashew butter over top as you like. Roll up the chicken as neatly as you can. Next, lay the strips of pancetta over the roast. Tie everything up with butcher's twine. Use one longer string running the length of the roast, and then tie every two inches in the other direction. Roast as directed for ballotine, above.

Duck Confit

This recipe and the next combine to allow you to cure a whole duck deliciously. Confit and rillettes—both French traditions—are two methods of curing meat in fat. I try to have one or both of them around all the time, or at least some leftover duck fat from the last confit (which is perfect for frying your morning eggs).

1 whole Pekin duck, about 5 pounds
1 whole corm garlic, plus 3 garlic cloves
2 Tbsp quatre epices
1½ oz. kosher salt
2 tsp ground rosemary

Remove all fat from duck and refrigerate in a covered container. Next, remove the whole legs; we'll confit these, and the rest of the bird will go to rillettes (see below). Combine the salt and quatre épices and rub thoroughly into the duck legs and duck carcass. Place the duck legs into a ziplock bag and place in the refrigerator. Peel the garlic cloves and mince them. Now rub the garlic and rosemary into the duck carcass. Place the seasoned carcass into its own ziplock bag and put in refrigerator.

After 24 hours, pull the duck legs out, rinse them thoroughly, and pat them dry. Place them in a small cast-iron skillet. Preheat the oven to 170°F–200°F. Next, slice the top off of the garlic corm and place the corm cut-side down in the cast-iron skillet, right beside the duck legs. Now remove the duck fat from the refrigerator and place it in the skillet so it covers the duck legs and garlic. If you don't have enough fat to cover, use some pork fatback to supplement.

Bring the skillet's contents to a simmer over medium heat, then transfer the skillet to the preheated oven and cook until fat is totally melted and clear and the legs are exceptionally tender and browned. Remove confit from oven and cool to room temperature, then place legs in a bowl and pour the oil over them. Use the garlic for something else—it will be deeply roasted in healthy seasoned fat and amazing.

Refrigerate the confit, covered, until it is well solidified. To serve, pull the legs from the fat and roast them in the oven until they are warm and crisp. You can save the fat almost indefinitely, in my experience.

Duck Rillettes

Ultimate cracker décor.

Duck carcass, rubbed as directed above, and refrigerated for 48 hours
½ bunch parsley
2 bay leaves
A few juniper berries
1½ lb. pork fatback, skin off, cut into 2 inch cubes
2 cups water

Preheat the oven to 250°F–300°F. Tie parsley, bay leaves and juniper berries into a cheesecloth sachet. Place duck carcass, fat back, water and sachet into an enameled cast-iron Dutch oven and bring to a simmer over medium–high heat. Cover, transfer to oven and braise, stirring every now and then, until the duck is super tender. This should take a couple of hours, if not more.

Remove from oven and cool slightly, then drain over a bowl to reserve the cooking liquid. Discard the herbs in cheesecloth. Next, pull the duck meat from the bone using your fingers, or a couple of forks if it is too hot. Mash the softened fatback and mix it with the shredded duck. Pack firmly into a pint mason jar. Pour as much of the reserved cooking liquid over top as you can. Cool to room temperature, place a lid on the jar, and chill completely in the refrigerator at least 1 day before serving. Serve the rillettes at room temperature.

Chicken Cardamom Sausage

This sausage is beautiful for breakfast, with citrus and eggs, or fancy for dinner with fried cauliflower, a drizzle of honey and a pinch of turmeric. Heirloom buckeyes are superior breeds for sausage making. Their dark meat and rich skin lends to beautifully colored sausage with lovely texture.

56 oz./3½ lb. boneless, skinless chicken, with as much fat on it as you can manage
16 oz./1 lb. chicken skin
1.5 oz. kosher salt
1.5 oz. brown sugar
0.7 oz. ground black pepper
0.7 oz. ground cardamom
0.5 oz. fresh garlic, grated on a microplane grater
0.4 oz. ground ginger
0.2 oz. ground nutmeg
0.4 oz. ground rosemary
0.2 oz. rubbed sage
½ cup full fat, plain yogurt
2 Tbsp sherry vinegar
2 Tbsp chicken stock
10 feet hog casings, rinsed and prepped

Cube up the chicken and reduce the skin to strips, laying all trim out on a rimmed baking sheet. Combine the yogurt, sherry vinegar and stock in a small bowl, then transfer to the refrigerator. In a separate bowl, combine remaining ingredients, then massage into the chicken trim thoroughly. Open-freeze the seasoned chicken trim for an hour or more, and be sure to put your grinder's moving parts in the freezer for at least 15–20 minutes before grinding.

When ready to grind, send all the chicken through the coarse grinder plate, then send one half of the ground mixture back through a second time. Next, add the yogurt mixture and mix thoroughly by hand. Test and adjust seasoning to your liking.

Stuff and refrigerate. Poach or pan cook and serve.

Sauces and Sundries For Poultry

Quick Buttermilk Drop Biscuits

2 cups all purpose flour, sifted for best results
1 Tbsp baking powder
2 tsp sea salt
¼ cup lard
¼ cup butter
1 cup buttermilk

Mix dry ingredients, then cut in cold lard and butter with your hands until they form pea-sized clumps. Dump in buttermilk and stir until just combined. Divide into 9–12 pieces and drop onto an ungreased baking sheet. Bake at 425°F for 10–12 minutes or until golden.

Milk Gravy

Start with fried chicken drippings or bacon grease. Melt as much of it in the pan as you like. For four people, I usually start with a tablespoon or so. To that, add a tablespoon of flour. Stir to create a smooth, light roux, then add whole milk in a slow stream, whisking to remove the clumps. Stop adding milk when the gravy reaches the consistency you're happy with. Season to taste with salt and pepper.

Pesto

Pesto is traditionally made with basil, but you can make it with anything. I make it with chickweed, arugula, collard stems, fennel fronds, dandelion leaves and more. If it's green and needs to be used, I make pesto. Freezes well.

1 cup basil or green leaves you need to use
¼ cup sunflower seeds, almonds, walnuts or pine nuts (whatever you have or whatever is cheapest)
2–3 cloves fresh garlic
¼ cup or more olive oil (depending on how thick you want the pesto)
¼ cup parmesan cheese (optional)
Salt and pepper to taste
Lemon juice (optional)

Combine all ingredients in a food processor and blend well.

Resources

Livestock Production Supplies

The best option is to look for local farm supply stores that carry fencing, watering and feeding equipment from larger manufacturers. If you can't find your needs with local business, try the following suppliers:
Premier1 Fences (premier1supplies.com)
Free Choice Minerals (freechoiceminerals.com)
Fertrell (fertrell.com)
FarmTek (farmtek.com)
Brower Poultry Equipment (browerequip.com)

Raising Cattle

Ranching Full-Time on Three Hours a Day by Cody Holmes, Acres, U.S.A., 2011.
How to Not go Broke Ranching by Walt Davis, self-published with CreateSpace, 2011.
Holistic Management Handbook: Healthy Land, Healthy Profits by Jody Butterfield, Sam Bingham and Allan Savory, Island Press, 2nd edition, 2006.
Grass-Fed Cattle by Julius Ruechel, Storey Publishing, 2006.
Salad Bar Beef by Joel Salatin, Polyface, 1996.
Nutrient Requirements for Beef Cattle, National Academies Press, 8th edition, 2015.
Management-Intensive Grazing by Jim Gerrish and Allan Nation, Green Park Press, 2004.
Comeback Farms by Greg Judy, Green Park Press, 2008.
Man, Cattle and Veld by Johann Zietsman, BEEFpower, 2014.
The New Livestock Farmer: The Business of Raising and Selling Ethical Meat by Rebecca Thistlewait, Chelsea Green, 2015

Raising Lamb

Storey's Guide to Raising Sheep by Carol Ekarius and Paula Simmons, Storey Publishing, 2009.
Sheep 101 & 201 (sheep101.info)

Raising Pork

Dirt Hog: A Hands-On Guide to Raising Pigs Outdoors... Naturally by Kelly Klober, Acres U.S.A., 2007.

Storey's Guide to Raising Pigs by Kelly Klober, Storey Publishing, 3rd edition, 2009.

Homegrown Pork: Humane, Healthful Techniques for Raising a Pig for Food by Sue Weaver, Storey Publishing, 2013.

Raising Poultry

The Small-Scale Poultry Flock by Harvey Ussery and Joel Salatin, Chelsea Green, 2011.

The Chicken Health Handbook and *Storey's Guide to Raising Chickens* by Gail Damerow, Storey Publishing, 2015 (5th edition) and 2010, respectively.

Raising the Home Duck Flock by Dave and Millie Holderread, Storey Publishing, 7th edition, 1990.

The Book of Geese by Dave Holderread, Hen House Publishing, 1993.

Chicken Tractor: The Gardener's Guide to Happy Hens and Healthy Soil by Andy Lee, Good Earth Publishing, 1994.

Homemade Living: Keeping Chickens by Ashley English, Homemade Living, 2010.

Resources For Butchery, Processing, etc.

Whole Beast Butchery by Ryan Farr and Brigit Binns, Chronicle Books, 2011.

The Art of Making Fermented Sausages by Stanley and Adam Marianski, Bookmagic, 2nd edition, 2012.

Charcuterie by Michael Ruhlman and Brian Polcyn, W.W. Norton and Co., 2013.

The Art of Charcuterie by John Kowalski and the Culinary Institute of America, Wiley, 2010.

Butchery and Sausage-Making for Dummies by Tia Harrison, For Dummies, 2011.

Butchering Beef and *Butchering Poultry, Rabbit, Lamb, Goat and Pork,* by Adam Danforth, Storey Publishing, 2014. These books cover slaughter, and do a very fine job of it.

The Butcher's Guild (thebutchersguild.org): a membership group of butchers, curers, chefs and meat enthusiasts

NC Choices (ncchoices.com): a program of NC State University working on niche meat supply chains. Even if you are not in NC, check out their work

Butcher & Packer (butcher-packer.com): for charcuterie and butchery supplies
Sausage Maker (sausagemaker.com)
LEM (lemproducts.com): processing equipment
Ultrasource (ultrasource.com): more for commercial scale, but good processing equipment

Thanks and Praise

To my incredible family: for knives and for rallying, for "I always knew it" and for fierce loyalty and support, for the way it should be. To my Big Dipper and my Lil' Dipper: forever & always. To New Society and Ingrid: for the opportunity. To Rob Hunt. To Jean-Martin: no funny business, OK? Your confidence and encouragement seriously lit a fire under me in this project. Thank you, thank you. To Shelby, sun in my eyes: you have been immense throughout this journey. I cannot thank you enough. To AJ, Matt & Lucy: my favorite people to love and to feed. To Cindy Kunst: let's get another project! You just a ray of light, girl. To Patryk Battle: thanks for your efforts, for your genius, your anarchy, your kindness and your frequent challenge. And thanks for the vote of confidence. I'll not forget it. To Joel DuFour: a true friend. To Rocco Sinicrope and Kendra Topalian Sinicrope: for jokes, and comfortable silence, and meat donations and the best pork I've ever cut. Y'all are quality, and hilarious. To my other colleagues at Living Web: your general support and encouragement has been a deep blessing. To Tim & Andrea: for lamb, and consistent goodness. For reminding me that everything can be healed. To Adam & Alyssa: for education, and inspiration.

To Gred Gross, Graham & Wendy Brugh, Brant & Shelly Bullock, Franny Tacy, Jim & Melissa Adkins. To Tanya Cauthen and Henry Reidy: y'all some badasses. And your meat thrift is like none I've ever seen. To Amy Price: thanks for letting me babble in your kitchen. Your shop is tip-top, congrats. To Kilan Brown: fired up, humble, and genuine. A true cutter. To Tyler Cook: you're a wizard, dude. To Matt Helms: love and solidarity always! To Asher & Julia: I'll always love y'all. Plus countless other people, for your encouragement and support.

Index

A

Achilles tendon, 73
Adkins, Jim, 200
Ager, Amy and Jamie, 10–11
aging, 25
aitch bone, 74, 107, 131
All Flesh is Grass, 46
all natural, 39
American Humane Association, 41
anchovy butter, 82
animal feeding operation (AFO), 36
Animal Welfare Approved, 39–40
animals, respect for, 1, 25
antibiotics, 38–39
apple butter, 150, 156
ascorbic acid (vitamin C), 185
aspic, 171
au gratin, 170
avian influenza, 206

B

backgrounding, 86
backslopping, 184
backstrap, 56
bacon
  beef bacon, 64, 79
  butchery, 135
  recipe, 178
bacteria
  in charcuterie, 158–159, 183, 184, 185
  importance of, xiii, 17
banh-mi sandwiches with quick pickles, 148
barbecue sauce, 155
barley, 47
baseball, 71, 72
basic pie crust, 151
basic salami, 186–187
bavette steak, 67, 136
beef
  breeds, 37, 42
  butchery. *see* beef butchery
  consumption of, 85
  cooking, 76–83. *see also* recipes
  feed, 44–48
  fencing, 43
  label claims, 38–41
  primals, 35
  raising, 36–37
  resources for, 225
  sauces and sundries, 81–82
  space and water, 42
beef and lovage sausage, 77
beef bacon, 64, 79
beef butchery. *see also individual cuts*
  chuck and brisket portion, 50–61
  forequarter, 49–50
  hindquarter, 65–66
  rib and plate portion, 61–65
  round, 73–76
beef jerky, 74, 80
beef stock, 76
beef tallow, 79
belly, pork, 141–142
Belmont Butchery, 65
black soldier fly (BSF), 202, 204
blackhead, 201
bologna, 176
bone marrow horseradish sauce, 81
bone saw, 30
bone scraper, 30
boning hook, 30
boning knife, 29
bookend, 141

bottom round, 74
bourbon and sorghum glazed lamb spare ribs, 114
bowtie, 134
braised beef shank tacos with caper chimichurri, 78
braised pork ribs with rooster sauce and balsamic vinegar, 149
bratwurst, 169
bread & butter pickles, 156
breakfast sausages, 168
breakfast scrapple with arugula, eggs and maple syrup, 153
breeds
  cattle, 37, 42
  pigs, 122–123
  poultry, 197–200
  sheep, 88–90
bresaola, 80
brining, 176–177
brisket, 50–51
broiled tomatillo salsa, 115
broody hens, 206
Brown, Kilan, 49
Brugh, Graham and Wendy, 20
Bullock, Brant and Shelly, 192
butcher's's knife, 29
butchery
  beef. *see* beef butchery
  lamb. *see* lamb butchery
  pork. *see* pork butchery
  poultry, 211–217
  resources for, 226–227
  tips, 28–32
  tools, 33
butt, pork, 143–145
buttermilk drop biscuits, 224
buying approaches, 4–12

C

campagne, 170
caper chimichurri, 78
capicola, 144, 181
carpaccio, 74
cashew cheese and chili pepper stuffing with pancetta, 221
casing, 32
casings, 166
Cauthen, Tanya, 65
celery juice powder, 163
certified humane raised and handled, 41
charcuterie
  categories, 158
  charcuterie chamber, 188–189
  described, 157
  dry curing, 176
  fundamentals of, 157–160
  humidity, 161
  nitrates, 161–164
  nitrites, 161–164
  pâtés, terrines, and meat specialties, 170–176
  pH, 161
  salt, 160
  sausage, fermented, 182–187
  sausage, fresh, 164–170
  skinning pork, 180
  smoke, 161, 178–182
  temperature, 160
  water activity, 160
  wet curing, 176–177
  whole-muscle cures, 176–178
cheesecloth, 32
chicharron with apple butter and cilantro crème fraîche, 150
chicken, 85. *see also* poultry
chicken ballotine, 220
chicken cardamom sausage, 223
chimichurri, 78
Chop Shop Butchery, 65, 95
chorizo, 168
chow chow, 147
chuck and brisket, 50–61
chuck eye roll, 56
cimeter, 29
citric acid, 185
cleaver, 30
Clostridium botulinum, 159, 162, 163
clothing, 31
cold smoking, 178

collar, 143
Concentrated Animal Feeding Operation (CAFO), 36
confit, 221
Cook, Tyler, 129
cooking, 12–15, 26–27. *see also* recipes; *specific meats*
corn, 47
corn pancakes, 147
Cornish Rock Cross, 191–192, 195–196
cost, 6–11, 20–23
coulotte, 71
creep feeding, 86
crème fraîche, 150
culatello, 132–133
cure #1, 161
cure #2, 162
curing. *see* charcuterie
curing salts, 161–164
cutting board, 31
cutting glove, 31

D

death, good, 1
deboning, 214–217
In Defense of Food, 15
dehasas, 126
Denver cuts, 55–56, 98
desertification, reversal of, 18–19
dextrose, 185
diet diversity, 15–20
dressed weight, 21
dressing, 25
dry curing, 176
drying, 167
duck confit, 221–222
duck rillettes, 222

E

Earl Grey braised lamb shank with herb dumplings, 110
eating variety, 15–20
electric fencing, 43
emulsions, 171
en croûte, 171
en terrine, 170
erythorbic acid, 185
ethical meat, defined, 1
external economy, 4
eye of round, 73–74

F

fat, 57, 68, 129
feed
  for cattle, 44–48
  for pigs, 127–128, 129
  for poultry, 201–204
  for sheep, 92–94
feed conversion ratio, 21
fencing, 43, 91–92, 127, 200–201, 203
fennel salami with nutmeg and wine, 187
fermenting, 182–187
filet mignon, 67
finished weight, 21
finishing ration, 40
fiochetto, 134, 182
fire-cooked lambchetta with apricot and rosemary, 112
first economy, 4, 19
flank, 66–67
flap meat, 66–67
flat iron, 57–59, 144
food, xiv–xv, 2
food Gestalt, 15
food processor, 32
food system, 2–3, 4–5
Foothills Deli and Butchery, 129
forequarter, 49–50
foreshank, 51–52
free choice, 40, 45
free range, 40–41
free roaming, 40–41
fried chicken, 219

G

Gaining Ground Farm, xiv–xv
garlic orange bratwurst, 169
Gdl, 185
genetically modified organisms, 3, 44
Gestalt, 15
ginger mint cilantro chutney, 115
glove, 31
GMO-free pork, 20–23
goats, 92, 93
good death, 1
good life, 1
goodness of food, xiv–xv
grains, 47
Grandin, Temple, 25
grasses, 45–46
grass-fed, 39, 45
gravy, milk, 224
grazing, 46, 90, 92–94
Grier, Anne, xiv–xv
grilled artichoke salad with smoked paprika aioli, 116
grinding, 166, 171–172, 174
grips, 29

H

Halpern, Matt, 65
ham hook, 30
Harrill, Walter, 219
headcheese, 175
Helms, Matt, 65, 95
herb dumplings, 110, 111
herbes de Provence sausages, 169
Hermetia illucens, 204
Hickory Nut Gap Farm (HNG), 10–11
high-tensile fencing, 43
hindquarter, 65–66
hip bone, 74
hogget, 87
Holistic Management, 46
honey mushroom duxelles stuffing, 220
honing steel, 29
hormones, 38
hot smoking, 178
hot vinegar sauce, 147
housing, for poultry, 200–201
humerus, 52–53
humidity, 161, 186

I

immersion blender, 32
inside round, 73
instacure #1, 161
invasive plants, 94–95

J

jerky, 74, 80

K

King Daddy's Chicken & Waffles, 219
King Family Farm, 192
knives, 28–30, 206
knuckle, 74–75
Kocuria, 183
kudzu, 94

L

label claims, 38–41
Label Rouge standard, 196
Lactobacillus, 183, 184
lamb
  breeds, 88–90
  butchery. *see* lamb butchery
  consumption of, 85
  cooking, 109–116. *see also* recipes
  economics of, 85–86
  feed, 92–94
  fencing, 91–92
  minerals, 93
  primals, 83
  raising, 85–88
  resources for, 225
  slaughter weight, 86, 90
  space and water, 90–91
lamb butchery. *see also individual cuts*
  chops, 98–99, 102–103
  leg, 106–109
  medallions, 104–105
  primals, 95, 95–97
  rib section, 100–103
  saddle, 97, 103–105
  shoulder, 98–100
lambchetta, 104
land, 117–120
lard, 151
lardo, 181
leaf fat, 129
legumes, 45–46
life, good, 1
lime-cream curry lamb sausage with dosas and raita, 111
liver pâté, 173–174
Livestock Conservancy, 198
livestock management, 18–19
livestock supplies, 225
Living Web Farms, 204
loaf pan, 32
Logsdon, Gene, 46
loin, 67–69, 103–104, 135–140
London broil, 73
Lundy, Ronni, 219

M

manure, 16–17
marmalade, 113
master brines, 177
master dry cure, 177
meat
  attitudes toward, 3
  ethical meat, 1
meat grinder, 31
meat hook, 30
meat thermometer, 26–27

Meatonomics, 6
middle economies, 4
middle market, 6
milk gravy, 224
minerals
  for cattle, 48
  for pigs, 128
  for poultry, 204
  for sheep, 93
mock tender, 54
molding, 172
mousseline, 171
MSG, 185
muscle, 26–27
mutton, 87

N
natural, in label claims, 39
netting and bags, 32
New York strip, 70
nitrates, 161–164, 183
nitrites, 161–164, 182, 183
no added hormones, 38
no antibiotics administered, 38–39

O
olive, feta and sundried-tomato stuffing, 220
onion, pickled red, 81
orange, fennel, and honey marmalade, 113
organic, 20–23, 41
osso bucco, 134
oyster, 74, 211–212

P
paddocks, for pigs, 124–126
panade, 171
pancetta stesa, 179
parasites, 89, 90
Pasteur, Louis, 159
pasture, 45–46, 48
pasture-raised, 40, 45
pâtés
  cooking, 172–176. *see also* recipes
  cooling, 173
  glossary, 170–171
  grinding, 171–172, 174
  molding, 172
  pâté de campagne, 174
  preparation, 171
Pediococcus, 183, 184
pellicle, 167
Pendulum Fine Meats, 49
pepperoni, 187
pesto, 224
petite tender, 59
pH, 161, 183
pH strips or meter, 32
Phillips, Michael, 17
pickled red onion, 81
pie crust, 151
pigs. *see* pork
pink salt, 161
plants, invasive, 94–95
plate primal, 64
plucking, 208
Pollan, Michael, 15
poly-wire fencing, 43
porchetta, 140–142, 154
pork
  breeds, 122–123
  butchery. *see* pork butchery
  cooking, 145–157. *see also* recipes
  cuts of, 117
  feed, 127–128, 129
  fencing, 127
  minerals, 128
  organic, GMO-free, 20–23
  raising pigs, 120–122
  resources for, 226
  skinning, 180
  space and water, 123–126
  U.S. pork industry, 120, 121
pork and pickle pie, 152
pork banh-mi sandwiches with quick pickles, 148
pork belly, 141–142
pork butchery. *see also individual cuts*
  belly, 135, 141–142
  butt, 143–145
  chops, 138
  leg primal, 131–134
  loin, 135–140
  primals, 128
  ribs, 138–139
  shoulder, 135, 142–145
porterhouse, 67, 105
potassium sorbate, 185
poultry
  additional considerations, 204–206

- breeds, 197–200
- broody hens, 206
- butchery, 211–217
- consumption of, 194
- cooking, 218–224. *see also* recipes
- deboning, 214–217
- feed, 201–204
- fencing, 200–201, 203
- home slaughter, 206–211
- housing, 200–201
- minerals, 204
- parts, 191
- pests and diseases, 204, 206
- raising, 194–197
- resources for, 226
- U.S. poultry industry, 195
- water, 202, 203, 204

prague powder, 161
predators, 91
price, 6–11, 20–23
primals, 49, 95–97, 128
prime rib, 62, 63
prosciutto, 180
pulled pork with hot vinegar sauce, chow chow, and corn pancakes, 147

Q

quatre epices, 174
queso fresco, 81–82
quick buttermilk drop biscuits, 224
quick pickles, 148

R

rail hook, 30
raita, 112
ranch roast, 59–60
rat, 75
recipes
- anchovy butter, 82
- apple butter, 156
- bacon, 178
- barbecue sauce, 155
- basic pie crust, 151
- basic salami, 186–187
- beef and lovage sausage, 77
- beef bacon, 79
- beef jerky, 80
- beef stock, 76
- beef tallow, 79
- bologna, 176
- bone marrow horseradish sauce, 81
- bourbon and sorghum glazed lamb spare ribs, 114
- braised beef shank tacos with caper chimichurri, 78
- braised pork ribs with rooster sauce and balsamic vinegar, 149
- bread & butter pickles, 156
- breakfast sausages, 168
- breakfast scrapple with arugula, eggs and maple syrup, 153
- bresaola, 80
- broiled tomatillo salsa, 115
- capicola, 181
- cashew cheese and chili pepper stuffing with pancetta, 221
- chicharron with apple butter and cilantro crème fraîche, 150
- chicken ballotine, 220
- chicken cardamom sausage, 223
- chorizo, 168
- cure recipes, 177
- dosas, 111
- duck confit, 221–222
- duck rillettes, 222
- Earl Grey braised lamb shank with herb dumplings, 110
- fennel salami with nutmeg and wine, 187
- fire-cooked lamb-chetta with apricot and rosemary, 112
- fried chicken, 219
- garlic orange bratwurst, 169
- ginger mint cilantro chutney, 115
- grilled artichoke

salad with smoked paprika aioli, 116
headcheese, 175
herb dumplings, 110
herbes de Provence sausages, 169
honey mushroom duxelles stuffing, 220
lard, 151
lardo, 181
lime-cream curry lamb sausage with dosas and raita, 111
liver pâté, 173–174
milk gravy, 224
olive, feta and sundried-tomato stuffing, 220
pancetta stesa, 179
pepperoni, 187
pesto, 224
pickled red onion, 81
porchetta with persimmon, chestnut and pine, 154
pork and pickle pie, 152
prosciutto, 180
pulled pork with hot vinegar sauce, chow chow, and corn pancakes, 147
quatre epices, 174
queso fresco, 81–82
quick buttermilk drop biscuits, 224
raita, 112
red wine mushrooms, 115
roast leg of lamb with mustard, capers, and marjoram, 114
roasted lamb rib with orange, fennel, and honey marmalade, 113
sauces and sundries, 81–82, 115–116, 155–156, 224
sauerkraut, 155
smoked fiochetto ham, 182
spatchcocked roasted chicken with lemon and basil, 218
red wine mushrooms, 115
resource cycles, 16–18
resources, 225–227
rib and plate portion, 61–65
rib primal, 62
ribeye, 63–64
rillettes, 221–222
roast leg of lamb with mustard, capers, and marjoram, 114
roasted lamb rib with orange, fennel, and honey marmalade, 113
rooster sauce, 149
roosters, 204
roosts, 200
rotational grazing, 46
the round, 73–76
rubber mallet, 30

S

saddle, 97, 103–105
salami, 186–187
salt, 160
sauces and sundries, 81–82, 115–116, 155–156, 224
sauerkraut, 155
sausage, fermented, 182–187
sausage, fresh
components of, 164–165
cooking, 167–170. *see also* recipes
drying, 167
grinding, 166
preparation, 165
problems with, 165
stuffing, 166–167
sausage stuffer, 31
Savory, Allan, 18, 45, 46, 95
scalding, 208
scales, 32
scapula, 54–55, 144
scrapple, 153
sheep. *see* lamb; mutton
shoes, 31
s-hook, 30
short loin, 69
short ribs, 64
shoulder, pork, 142–145

shoulder clod, 53–54, 57–59
silvopasture, 124
Simon, David Robinson, 6
sirloin, 69–72, 131
skinning pork, 180
skirt, 61, 135
slaughter
  general notes, 25–26
  lamb slaughter weight, 86
  of poultry, 206–211
slaughter weight, 90
smearing, 166
smoked fiochetto ham, 182
smoked paprika aioli, 116
smoking meats, 161, 178–182
sodium nitrite, 161, 162
soy protein, 185
space
  for cattle, 42
  for pigs, 123–126
  for poultry, 200–201
  for sheep, 90–91
spatchcocked roasted chicken with lemon and basil, 218
spice grinder, 32
spider, 74, 132
splenius, 55
sprouting grains, 128, 129, 202
Staphylococcus, 183
starter culture, 185
STC-1000, 189
storage supplies, 31
strip loin, 140
stuffing, 166–167
subsidies, 6, 7
succession, 94–95
suet, 68
supplemental feed, 39, 40, 45, 47
Sustainable Poultry Network, 198
systems, in nature, 15–18

T

tacos, 78
tallow, 79
T-bone, 67, 105
TCM, 161
temperature, 27, 160
tenderloin, 67–68, 69, 104, 105, 129
terrine mold, 32
terrines, 170
terroir, 126
time, 14
tools, 28–33
top blade, 57–59
top round, 73
top sirloin center, 71
trap crops, 204
tri-tip, 68–69
trotter, 134

U

underarm roast, 59–60
USDA certified organic, 41

V

Vegas steak, 55
vertical integration, 6, 7
vinegar sauce, 147

W

warm smoking, 178
waste, 121–122
water
  for cattle, 42
  for pigs, 126
  for poultry, 203, 204
  for sheep, 90–91
  usage, 16–17
water activity, 160
wet curing, 176–177
WH8040, 189
whole-muscle cures, 176–178
worktable, 30

About the Author

Over the past thirteen years, MEREDITH LEIGH has worked as a farmer, chef, teacher, non-profit executive director, and writer, all in pursuit of sustainable food. She has developed a farmers cooperative, catalyzed non-profit farm ventures, raised vegetables, flowers, and pastured meats, owned and managed a retail butcher shop, and more. Currently, Meredith spends some of her time teaching, handling sheep, cooking, and doing outreach at Living Web Farms, a non-profit education and research farm in Mills River, NC. She is a mother to two boys, many plants and fermentation projects, and lives in Asheville, NC.

New Society Publishers

ENVIRONMENTAL BENEFITS STATEMENT

New Society Publishers has chosen to produce this book on recycled paper made with 100% post consumer waste, processed chlorine free, and old growth free.

For every 5,000 books printed, New Society saves the following resources:[1]

34	Trees
3,055	Pounds of Solid Waste
3,361	Gallons of Water
4,384	Kilowatt Hours of Electricity
5,554	Pounds of Greenhouse Gases
24	Pounds of HAPs, VOCs, and AOX Combined
8	Cubic Yards of Landfill Space

[1]Environmental benefits are calculated based on research done by the Environmental Defense Fund and other members of the Paper Task Force who study the environmental impacts of the paper industry.